Drugs Affecting the Renin-Angiotensin-Aldosterone System

Progress in Biochemical Pharmacology

Vol. 12

Series Editor: R. Paoletti, Milan

S. Karger · Basel · München · Paris · London · New York · Sydney

Proceedings of the 5th Kanematsu Conference on the Kidney, Sydney 1975

Drugs Affecting the Renin-Angiotensin-Aldosterone System

Use of Angiotensin Inhibitors

Volume Editors
G. S. STOKES, Sydney, and K. D. G. EDWARDS, New York, N.Y.
Kanematsu Memorial Institute, Sydney Hospital, Sydney and
Clinical Physiology and Renal Service, Memorial Sloan Kettering Cancer Center, New York, N.Y.

98 figures and 18 tables, 1976

S. Karger · Basel · München · Paris · London · New York · Sydney

Progress in Biochemical Pharmacology

Vol. 9: Drugs and the Kidney. Editor: EDWARDS, K. D. G. (Sydney). XII + 274 p. 64 fig., 43 tab., 1974
ISBN 3-8055-1693-2

Vol. 10: Lipids and Tumors. Editor: CARROLL, K. K. (London, Ont.). X + 400 p., 32 fig., 60 tab., 1974
ISBN 3-8055-1708-4

Vol. 11. Biological Basis of Clinical Effect of Bleomycin. Editor: CAPUTO, A. (Rome). X + 230 p., 74 fig., 69 tab., 1976
ISBN 3-8055-2338-6

Cataloging in Publication
    Kanematsu Conference on the Kidney, 5th, Sydney, 1975
        Drugs affecting the renin-angiotensin-aldosterone system: use of angiotensin inhibitors; proceedings
        Volume editors G. S. Stokes and K. D. G. Edwards. Basel, New York, Karger, 1976.
        (Progress in biochemical pharmacology, v. 12)
        1. Hypertension – diagnosis – congresses 2. Renin – metabolism – congresses 3. Angiotensin antagonists & inhibitors – congresses 4. Aldosterone – metabolism – congresses I. Stokes, G. S., ed. II. Edwards, K. David G., ed. III. Title IV. Series
        W1 PR666H v. 12/WK180 K16 1975d
        ISBN 3-8055-2410-2

All rights, including that of translation into other languages, reserved.
Photomechanic reproduction (photocopy, microcopy) of this book or parts thereof without special permission of the publishers is prohibited.

© Copyright 1976 by S. Karger AG, Basel (Switzerland), Arnold-Böcklin-Strasse 25
Printed in Switzerland by Buchdruckerei National-Zeitung AG, Basel
ISBN 3-8055-2410-2

Contents

Introduction XIII

Section I: General Reviews

Angiotensin II Blockade and the Functions of the Renin-Angiotensin System
J. O. DAVIS, R. H. FREEMAN, B. E. WATKINS, G. A. STEPHENS and G. M. WILLIAMS, Columbia, Mo. 1

Introduction 1
Effects of an Angiotensin II Antagonist on Aldosterone Secretion in the Dog 2
Effect of Angiotensin II Blockade on Aldosterone Secretion in the Rat 5
Angiotensin II Blockade in Experimental High Output Failure and in Experimental Renovascular Hypertension 9
Summary 14
References 14

Experimental and Clinical Studies with Converting Enzyme Inhibitor
E. HABER and A. C. BARGER, Boston, Mass. 16

Introduction 16
Renin and Sodium Balance 17
Studies in Normal Human Subjects 20
Experimental Renovascular Hypertension 22
Renin-Dependent Hypertension in Human Subjects 26
Conclusion 29
References 29

Section II: Studies in Experimental Animals

Comparative Studies of the Humoral and Arterial Pressure Responses to Sar^1-Ala^8-, Sar^1-Ile^8- and Sar^1-Thr^8-Angiotensin II in the Trained, Unanaesthetized Dog
E. L. Bravo, M. C. Khosla and F. M. Bumpus, Cleveland, Ohio 33

Introduction 33
Methods 34
Results 35
    Effect on Arterial Blood Pressure 35
    Effect on PRA 38
    Effect on Plasma Aldosterone 38
Discussion 38
Summary 39
References 39

Stimulating Effects of Angiotensin I, Angiotensin II and des-Asp^1-Angiotensin II on Steroid Production in vitro and its Inhibition by Sar^1-Ala^8- Angiotensin II
R. Hepp, C. Grillet, A. Peytremann and M. B. Vallotton, Geneva 41

Introduction 41
Methods 42
Results 42
Discussion 45
Summary 46
References 47

Discussion 49

Effect of Adrenal Arterial Infusion of Saralasin (P113) on Aldosterone Secretion
J. R. Blair-West, J. P. Coghlan, D. A. Denton, H. D. Niall, B. A. Scoggins and G. W. Tregear, Parkville, Vict. 53

Introduction 53
Results and Discussion 54
Conclusions 61
Summary 61
References 61

Effects of Saralasin on Renal Function in the Rat
K. G. Hofbauer, K. Bauereiss, H. Zschiedrich and F. Gross, Heidelberg ... 63

Introduction ... 63
Materials and Methods ... 64
    Isolated Perfused Rat Kidney ... 64
    Acute Renal Failure ... 65
    Substances ... 65
    Statistical Analysis ... 65
Experiments ... 66
    Isolated Perfused Rat Kidney ... 66
    Acute Renal Failure ... 67
Results ... 68
    Isolated Perfused Rat Kidney ... 68
    Acute Renal Failure ... 70
Discussion ... 71
    Intrinsic Activity of AII Antagonists ... 71
    Renal Haemodynamics and Circulating AII ... 72
    Intrarenal Formation of AII ... 73
    Autoregulation and Intrarenal AII ... 74
    Renin Release ... 74
    Renal Prostaglandins ... 75
    Acute Renal Failure ... 75
Summary ... 76
Acknowledgements ... 77
References ... 77

Discussion ... 84

Effects of Angiotensin Antagonists in Various Forms of Experimental Arterial Hypertension
C. M. Ferrario, F. M. Bumpus, Z. Masaki, M. C. Khosla and J. W. McCubbin, Cleveland, Ohio ... 86

Introduction ... 86
Methods ... 87
Results ... 88
    Effect of Angiotensin Antagonists on Renovascular Hypertension ... 88
    Findings in Chronic Hypertension Due to Cellophane Perinephritis ... 90
    Effect of Angiotensin Blockers in Malignant Hypertension ... 91
Discussion ... 92
Summary ... 95
Acknowledgements ... 95
References ... 96

Antagonists, Inhibitors and Antisera in the Evaluation of Vascular Renin Activity
The Role of Local Generation of Angiotensin II
J. D. SWALES, Leicester 98

Introduction 98
Vascular Wall Renin Activity 99
Changes in Vascular Renin 99
Role of Vascular Renin Activity 100
Vascular Renin and Angiotensin Antisera 101
Pressor Responsiveness to AII and Sodium Balance 104
AII Antagonist versus Antiserum 105
Timing of Pressor Changes after Nephrectomy 108
Response to AII Antagonist and Inhibitor after Nephrectomy 109
Conclusions 110
Summary 111
References 111

Discussion 114

The Use of Saralasin to Evaluate the Function of the Brain Renin-Angiotensin System
I. A. REID, San Francisco, Calif. 117

Introduction 117
Effects of Central Administration of Renin 119
  Drinking 119
  Blood Pressure 120
  ADH Secretion 122
Effects of Central Administration of Substrate 124
Effects of Central Administration of Agents which Block the Renin-Angiotensin System 125
  Drinking 126
  Blood Pressure 127
  Possible Role of Centrally Generated Angiotensin II in Hypertension 129
  ADH Secretion 130
Concluding Remarks 130
Summary 131
References 132

Competitive Inhibitors of Renin
A Review
K. POULSEN, Copenhagen, J. BURTON and E. HABER, Boston, Mass. 135

Introduction 135
Results and Discussion 136

Summary 140
References 140

Discussion 142

Section III: Clinical Studies

Angiotensin II Blockade in Normal Man and Patients with Essential Hypertension. Blood Pressure Effects Depending on Renin and Sodium Balance
H. R. BRUNNER, Lausanne, H. GAVRAS, Boston, Mass., A. B. RIBEIRO and
L. POSTERNAK, New York, N.Y. 145

Introduction 145
Methods 146
    Normal Volunteers 146
    Patients with Essential Hypertension 147
    Laboratory Determinations 149
Results 149
    Normotensive Volunteers 149
        Blood Pressure 151
        Sodium Balance 151
        Plasma Renin Activity 151
    Patients with Essential Hypertension 152
        Blood Pressure 154
        Sodium Balance 157
        Plasma Renin Activity 157
Discussion 157
Summary 160
References 161

The Role of Renin in the Control of Blood Pressure in Normotensive Man
P. J. MULROW and R. NOTH, New Haven, Conn. 163

Introduction 163
Methods 163
Results 166
    Normal Subjects 166
    Cirrhotic Patients 167
Discussion 167
Summary 169

References 169

Discussion 170

Changes of Blood Pressure, Plasma Renin Activity and Plasma Aldosterone Concentration following the Infusion of Sar[1]-Ile[8]-Angiotensin II in Hypertensive, Fluid and Electrolyte Disorders
T. YAMAMOTO, K. DOI, T. OGIHARA, K. ICHIHARA, T. HATA and Y. KUMAHARA, Osaka 174

Introduction 174
Materials and Methods 176
Results 177
    Normal Subjects in Three Different States of Sodium Balance 177
    Patients with Hypertension, Fluid and Electrolyte Disorders 178
    Relation between BP Changes and Pre-Infusion PRA 183
    Relation between PAC Changes and Pre-Infusion PRA 183
    Relationships between Changes in MBP, PRA and PAC 183
Discussion 183
Summary 186
References 187

Angiotensin II Blockade in Hypertensive Dialysis Patients
G. A. MACGREGOR, London, and P. M. DAWES, Alderley Park, Cheshire 190

Introduction 190
Methods 191
Case Reports 192
Discussion 196
Summary 198
Acknowledgements 198
References 198

Discussion 200

Angiotensin Antagonists as Diagnostic and Pharmacologic Tools
W. A. PETTINGER and H. C. MITCHELL, Dallas, Texas 203

Introduction 203
Pharmacokinetics of Saralasin 205
Diagnostic Use of Saralasin 206

Hypotensive Response	208
Augmentation of Split Vein Renins	209
Autonomous Aldosterone Secretion	210
Angiotensin Antagonists as Tools in Clinical Pharmacology	211
Summary	212
References	212

The Use of Saralasin in the Recognition of Angiotensinogenic Hypertension
D. H. P. STREETEN, T. G. DALAKOS and G. H. ANDERSON, jr., Syracuse, N.Y.. 214

Introduction	214
Method of Use of Saralasin	216
Need for Antecedent Na Loss	216
Method of Administration of Saralasin	218
Significance of Hypotensive Response to Saralasin	219
Nature of Renal and Renovascular Lesions in Saralasin Responders	222
Summary	224
References	224

Discussion 227

The Effects of the Angiotensin II Antagonist Saralasin on Blood Pressure and Plasma Aldosterone in Man in Relation to the Prevailing Plasma Angiotensin II Concentration
J. J. BROWN, W. C. B. BROWN, R. FRASER, A. F. LEVER, J. J. MORTON, J. I. S. ROBERTSON, E. A. ROSEI and P. M. TRUST, Glasgow, U.K. 230

Introduction	230
Methods	231
Results	233
Normal Subjects	233
Essential Hypertension	234
Primary Aldosteronism	234
Hypertension with Severe Chronic Renal Failure	234
Hypertension with Unilateral Renal Artery Stenosis	236
Relationship between Basal Plasma Angiotensin II Concentration and Blood Pressure Change during Saralasin Infusion	236
Relationship between Basal Plasma Angiotensin II Concentration and Change in Plasma Aldosterone during Saralasin Infusion	237
Discussion	237
Summary	239
Acknowledgements	240
References	240

Haemodynamic Effects of Sar¹-Ala⁸-Angiotensin II in Patients with Renovascular Hypertension
R. FAGARD, A. AMERY, P. LIJNEN, T. REYBROUCK and L. BILLIET, Leuven 242

Introduction 242
Methods... 243
Results 244
    Characteristics of the Patients 244
    Mean Arterial Pressure and Heart Rate in Sodium-Replete Patients 244
    Mean Arterial Pressure and Haemodynamic Variables in Sodium-Deplete Patients.. 245
    Mean Arterial Pressure in Patients Studied before and after Sodium Depletion 246
Discussion... 247
Summary.. 248
References... 248

Discussion 250

Subject Index 252

Introduction

This volume contains a collection of recent work on the use of angiotensin inhibitors in the diagnosis of human hypertension, and in analysing the pathophysiology of certain experimental situations, such as sodium depletion, high output heart failure, acute renal failure or renal-clip hypertension.

Theoretically, the actions of angiotension could be inhibited by preventing its binding to receptors, by interfering at any stage in its production from renin substrate or by interfering with the production or actions of its target hormones, the most important of which is aldosterone. Numerous potential modes of inhibition exist. Interference with the production of renin substrate from the liver or of renin from the kidneys, competitive inhibition of renin or of the enzyme which converts the decapeptide angiotensin I to angiotensin II, and antagonists or antibodies to the effector angiotensins 'II' (the octapeptide) and 'III' (the heptapeptide) have all been described, as have inhibitors of aldosterone biosynthesis and action.

The book does not attempt to deal in a sequential way with the entire subject of inhibition of the renin-angiotensin system. Some topics which are adequately reviewed elsewhere, such as the blockade of renin release and the role of aldosterone antagonists, are not mentioned, while others, namely competitive inhibition of renin and inhibition of converting enzyme, are each the subject of a single special chapter. The principle thrust of this presentation is to explore the practical application of angiotensin analogues, which have proved the most specific type of angiotensin inhibitor yet devised for clinical use. The analogue which has had the widest use is the octapeptide saralasin ('P113' – Norwich Pharmacal Company) in which sarcosine has been substituted for aspartic acid at the N-terminal end of the angiotensin II molecule and alanine for phenylalanine at the C-terminal end. The abbreviated chemical nomenclature we have used for this peptide is Sar^1-Ala^8-angiotensin II. Other analogues discussed are Sar^1-Ile^8-angiotensin II and Sar^1-Thr^8-angiotensin II.

The contributing authors, all of whom have been prominent in recent developments in this field, were originally invited to present their papers at the Fifth Kanematsu Conference on the Kidney. This conference, which was held at Sydney Hospital in February 1976 as a satellite of the Fourth Meeting of the International Society of Hypertension, had as its topic *The Use of Angiotensin Inhibitors in Clinical Diagnosis*. The introductory reviews in section I were contributed by Prof. J. O. DAVIS, who delivered the 1975 Volhard Lecture of the International Society of Hypertension on the subject of blocking agents and the renin-angiotensin system, and by Prof. EDGAR HABER, whose studies with the nonapeptide converting enzyme inhibitor in normal subjects have paved the way for understanding the role of the renin-angiotensin system in the adjustments to posture and sodium depletion in man. Both these authors graciously submitted manuscripts even though they were unable to come to Australia for the Conference. Sections II and III, covering experimental and clinical research, respectively, contain the 16 papers which were read at the Conference, together with the transcripts of the discussion periods which followed each pair of papers.

In the task of organising the Fifth Kanematsu Conference and editing the papers and transcripts, we have had unstinting support and much practical help from the members of the Cardio-Renal Unit, Sydney Hospital. We owe a particular debt of gratitude to Mr. IAN THORNELL, Dr. HELEN OATES and Miss JUDY GAIN.

G. S. STOKES
K. D. G. EDWARDS

Section I: General Reviews

In STOKES and EDWARDS: Drugs Affecting the Renin-Angiotensin-Aldosterone System. Use of Angiotensin Inhibitors
Prog. biochem. Pharmacol., vol. 12, pp. 1–15 (Karger, Basel 1976)

Angiotensin II Blockade and the Functions of the Renin-Angiotensin System

JAMES O. DAVIS, RONALD H. FREEMAN, BARRY E. WATKINS, GREGORY A. STEPHENS and GARY M. WILLIAMS

Department of Physiology, University of Missouri School of Medicine, Columbia, Mo.

Contents

Introduction 1
Effects of an Angiotensin II Antagonist on Aldosterone Secretion in the Dog 2
Effect of Angiotensin II Blockade on Aldosterone Secretion in the Rat 4
Angiotensin II Blockade in Experimental High Output Failure and in Experimental Renovascular Hypertension 9
Summary 14
References 14

Introduction

In 1960–62, a large body of evidence accumulated to show that the renin-angiotensin system is a primary controller of aldosterone secretion [1]. Most of the early evidence came from studies in dogs and man and attempts to extend these findings to other species, especially the rat, failed to show a clear-cut relationship. Also, many of the observations were made during sodium depletion, which is a potent stimulus for aldosterone secretion. BLAIR-WEST et al. [2] have done extensive studies on the mechanisms controlling aldosterone secretion during sodium depletion in sheep and reported that factors other than the renin-angiotensin system, ACTH and plasma electrolyte concentrations are involved. The present report includes extensive studies of the mechanisms controlling adrenal steroid secretion, during sodium depletion in the dog and rat, and during thoracic caval constriction in the dog, by use of angiotensin II blockade. In addition, data are presented on the role of the renin-angiotensin system in experimental high output failure and in experimental renovascular hypertension.

Effects of an Angiotensin II Antagonist on Aldosterone Secretion in the Dog

The angiotensin II analogue which has been used most extensively and most successfully for angiotensin II blockade is 1-sarcosine-8-alanine angiotensin II. Since most of the controversy in the past has centered around the control mechanisms for aldosterone secretion during sodium depletion, our most extensive studies have been directed at this specific problem by use of Sar^1-Ala^8-angiotensin II. However, even before blocking agents were available, the effects of bilateral nephrectomy were studied in sodium-depleted hypophysectomized dogs [3]. Aldosterone secretion fell to very low levels following nephrectomy and corticosterone production which was very low as a result of hypophysectomy fell further after nephrectomy (fig. 1). The experiments were conducted in hypophysectomized dogs to prevent a high level of ACTH secondary to laparotomy from obscuring a possible fall in aldosterone secretion after removal of the kidneys. When the problem of the control of aldosterone secretion during sodium depletion was reinvestigated recently by use of the angiotensin II antagonist, Sar^1-Ala^8-angiotensin II [4], aldosterone secretion fell to levels indistinguishable from zero in 5 of 7 dogs (table I); in these studies, dexamethasone was given to depress ACTH release. Interesting incidental findings were the striking increase in plasma renin activity (PRA) and the decrease in arterial pressure. As suggested elsewhere [5], it seems likely that the increase in PRA resulted from interruption of the negative feedback of angiotensin II on the JG cells and from the fall in arterial pressure. The studies were conducted in anaesthetized dogs with a chronic indwelling catheter for collection of adrenal venous blood and for direct measurement of the rate of aldosterone secretion.

In recent unpublished observations, this experiment was repeated with measurements of the concentration of plasma aldosterone by radioimmunoassay in conscious sodium-depleted dogs. Sar^1-Ala^8-angiotensin II was given for prolonged periods up to 210 min. A striking fall to the extent of 73–92% in the plasma aldosterone level occurred and

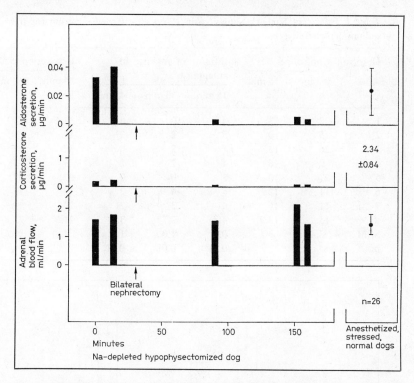

Fig. 1. Effects of bilateral nephrectomy in sodium-depleted hypophysectomized dogs. For comparison, data are presented for anesthetized, stressed, normal dogs. Reprinted with permission of the *Journal of Clinical Investigation* [3].

was sustained in 2 of the 3 dogs throughout the period of infusion of the angiotensin II antagonist. In the third animal, plasma aldosterone decreased initially and then increased from this very low level during the last hour of analogue infusion. Under these circumstances, the presence of an increased plasma level of angiotensin II might have produced a decrease in the efficacy of Sar^1-Ala^8-angiotensin II in the maintenance of a low plasma level of aldosterone; an increase in PRA did occur in all 3 animals. The expected progressive fall in arterial pressure also occurred. Conscious normal dogs studied similarly failed to show a change in either PRA or in arterial pressure.

The observations during sodium-depletion were extended to another experimental model in the dog, namely chronic constriction of the thoracic inferior vena cava. The results in this model have important clinical

Table I. Effects of Sar^1-Ala^8-angiotensin II on aldosterone secretion (ng/min) in sodium-depleted dogs

Dog No.	Control periods		Infusion of angiotensin antagonist			Recovery periods	
	0 min	15 min	15 min	30 min	45 min	45 min	60 min
1	22	23	4	0	8	33	19
2	5	10	5	2	2	11	–
3	26	13	7	8	6	32	57
4	9	6	4	2	2	3	2
5	27	10	3	0	0	27	27
6	11	12	6	4	3	17	13
7	11	16	4	3	2	20	19
$\bar{x}$	15.9	12.9	4.7	2.7	3.3	21.3	22.8
SEM	3.4	3.1	0.5	1.0	1.0	4.3	7.6
p			<0.01	<0.01	<0.01		

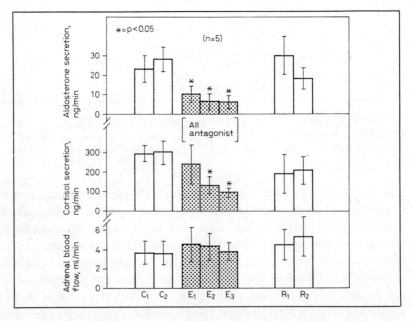

Fig. 2. Effects of intravenous infusion of the angiotensin II antagonist, Sar^1-Ala^8-angiotensin II on steroid secretion in dogs with chronic thoracic inferior vena caval constriction. The abreviations C_1, C_2 are for control periods and R_1, R_2 are for recovery periods. The three experimental periods, E_1, E_2, and E_3, are for measurements made at 15, 30, and 45 min of infusion of the angiotensin II antagonist. Reprinted with permission of *Science* [4].

implications because the alterations in salt and water metabolism simulate closely those occurring in patients with low output congestive heart failure. Again, Sar[1]-Ala[8]-angiotensin II produced a striking fall in aldosterone secretion in anaesthetized dogs (fig. 2). An important incidental finding was a striking progressive decrease in cortisol secretion. Since cortisol is secreted by the zona fasciculata and zona reticularis, these findings demonstrate that angiotensin II receptors are present in the two inner zones of the adrenal cortex. An increase in PRA occurred and arterial pressure fell, similar to the changes observed during infusion of this angiotensin II antagonist into sodium-depleted dogs.

In recent unpublished studies, one conscious dog with thoracic caval constriction was given Sar[1]-Ala[8]-angiotensin II for 210 min. Plasma aldosterone concentration decreased from 100 to 17 ng% while PRA increased from an elevated level of 30 ng angiotensin II/ml plasma to 144 ng angiotensin II/ml plasma. Arterial pressure fell from 100 to 70 mm Hg.

Effect of Angiotensin II Blockade on Aldosterone Secretion in the Rat

As indicated above, the role of the renin-angiotensin system in the regulation of aldosterone secretion in the rat has been difficult to define. Most of the studies have been made in sodium-depleted animals. SPIELMAN and DAVIS [6] have recently reported a detailed study on the mechanisms controlling aldosterone secretion in the sodium-depleted rat. Sodium depletion was achieved by reducing sodium intake from a normal control level of 2.89 to 0.04 mEq/day for 7–10 days; the average negative sodium balance was 1.90 mEq. Under these circumstances, aldosterone secretion increased from a normal rate of 0.66 to 7.70 ng/min. Simultaneous measurements of corticosterone secretion revealed that corticosterone output was high in both the normal and sodium-depleted rats due to the stress of laparotomy; a large dose of dexamethasone was given 2 h before the collection of adrenal venous blood but this failed to block ACTH release.

It is well known that in the rat sodium depletion leads to a rise in PRA and a striking increase in aldosterone secretion raising the question of causal relationship. To answer this question, the nonapeptide converting enzyme inhibitor (SQ 20881) was given to 3 groups of sodium-deplet-

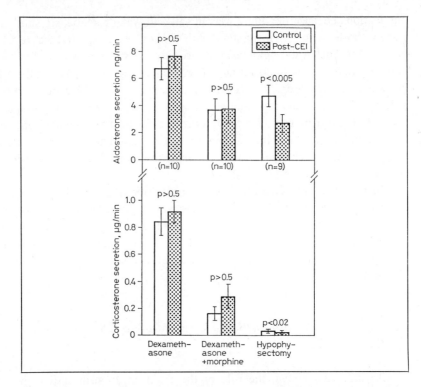

Fig. 3. Effects of a nonapeptide converting enzyme inhibitor (SQ 20881) on steroid secretion in three groups of sodium-depleted rats. Control values are unshaded columns while stippled columns are for values obtained after administration of the converting enzyme inhibitor (CEI). Reprinted with permission of the American Heart Association [6].

ed rats (fig. 3). In the 3 groups of animals, anterior pituitary function was high, at an intermediate level, or at a very low level as a result of hypophysectomy. These 3 levels of anterior pituitary function were used because in the preliminary study [6] the findings suggested that an elevated plasma level of ACTH secondary to the stress of laparotomy might obscure the fall in aldosterone secretion during angiotensin II blockade. As indicated by the control rates of corticosterone secretion (open bars), dexamethasone alone failed to suppress ACTH release (group a), dexamethasone and morphine gave partial suppression (group b), and hypophysectomy reduced corticosterone secretion to a very low level (group c). Under the conditions of elevated ACTH in groups a and b, angiotensin II blockade failed to show a decrease in aldosterone secretion. When, however, ACTH was excluded by hypophysectomy (group 3), a striking drop in aldosterone secretion occured during blockade with the converting enzyme inhibitor. Also, corticosterone secretion decreased from a very low level of 33 ng/min after hypophysectomy to 22 ng/min during angiotensin II blockade ($p<0.02$). This decline in corticosterone production during angiotensin II inhibitor agrees with the observation that angiotensin II increases the rate of corticosterone secretion in the rat as it does in other mammalian species [1]. It is of interest that arterial pressure fell significantly in all 3 groups of sodium-depleted rats. These observations with the converting enzyme inhibitor demonstrated a role for the renin-angiotensin system in the control of aldosterone secretion in the rat, but it was necessary to exclude the obscuring influence of ACTH for this demonstration.

More recently, angiotensin II blockade with Sar^1-Ala^8-angiotensin II in hypophysectomized, sodium-depleted rats [7] has confirmed the observation with the converting enzyme inhibitor. The steroid response to this angiotensin II antagonist was striking (fig. 4). Aldosterone secretion which was increased fourfold by sodium depletion fell 80% and, again, corticosterone secretion which was very low (35 ng/min) due to hypophysectomy also fell markedly. A 50% drop in arterial pressure occurred. These observations add support to the previous results which reveal an important role for the renin-angiotensin system in the control of aldosterone biosynthesis in the rat.

With the realization that a high plasma level of ACTH can obscure a steroid response during angiotensin II blockade, it became apparent that under ideal circumstances intravenous angiotensin II administration might produce a larger increase in aldosterone than that obtained in the early studies. Consequently, the response to intravenous infusion of synthetic angiotensin II was studied in rats with ACTH suppressed by administration of dexamethasone and morphine and this was compared to the earlier response with dexamethasone alone [8]. During pretreatment with dexamethasone and morphine, an infusion of 1 ng/min of angiotensin II increased aldosterone secretion from 0.53 ± 0.22 SEM ng/min to 2.62 ± 0.70 ng/min ($p<0.005$). Corticosterone secretion also increased from 145 ± 271 ng/min ($p<0.05$). In contrast, when ACTH was high initially from the stress of laparotomy during dexamethasone alone the aldosterone response was barely discernible. These results agree with the marked increases in plasma aldosterone concentration observed by COLE-

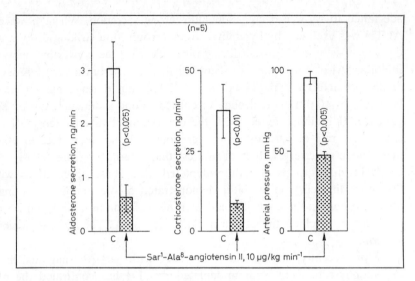

Fig. 4. Effects of Sar[1]-Ala[8]-angiotensin II in hypophysectomized sodium-depleted rats on steroid secretion and arterial pressure. Reprinted with permission of *Federation Proceedings* [7].

MAN *et al.* [9] and by CAMPBELL and PETTINGER [10] in response to angiotensin II given to conscious rats, which were decapitated for collection of blood.

Evidence for the role of the renin-angiotensin system in the control of aldosterone secretion has also been provided from studies of the heptapeptide fragment of angiotensin II [8]. When Des-Asp[1]-angiotensin II (1 μg/min) was given to rats pretreated with dexamethasone and morphine, aldosterone secretion increased from 0.55 ± 0.16 to 1.30 ± 0.28 ng/min (p<0.02). Sar[1]-Ala[8]-angiotensin II (50 μg/kg min^{-1}) blocked the aldosterone response to simultaneously infused Des-Asp[1]-angiotensin II (1 μg/min), but a much smaller dose of the angiotensin II antagonist was required for blockade of the same dose of angiotensin II. These findings suggest that the heptapeptide has a higher affinity for zona glomerulosa receptors than the octapeptide. These *in vivo* observations in the rat are consistent with the recent suggestion of CHIU and PEACH [11], from *in vitro* studies of aldosterone biosynthesis, that the heptapeptide might be a physiological mediator of the renin-angiotensin system. The recent finding of SEMPLE and co-workers [personal commun.] that the peripheral plasma level of the heptapeptide is equal to that of angiotensin

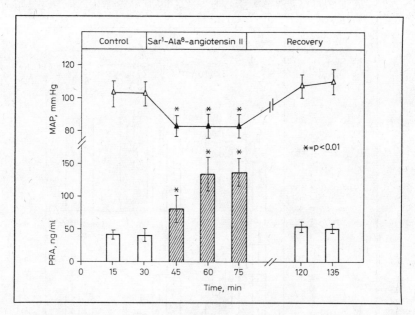

Fig. 5. The responses in mean arterial pressure (MAP) and plasma renin activity (PRA) in conscious dogs with high output failure to the intravenous infusion of Sar^1-Ala^8-angiotensin II at 6.0 µg/kg/min. MAP was measured in the femoral artery below the aortic-caval fistula. Values are means ± SEM, n = 5. Reprinted with permission of the *American Journal of Physiology* [12].

II in the rat gives credibility to the hypothesis that at least in this species the Des-Asp^1-angiotensin II mediates the response in aldosterone secretion.

Angiotensin II Blockade in Experimental High Output Failure and in Experimental Renovascular Hypertension

Experimental high output failure in dogs was produced by placement of a large aortic-caval fistula [12]. This model is characterized by an elevation in mean right atrial pressure, hepatomegaly, marked sodium retention, oedema and ascites. Renal blood flow is reduced, renin secretion and PRA are increased and hyperaldosteronism occurs. As the syndrome progresses, ventricular failure extends from the right to the left side of the heart and death usually occurs from pulmonary oedema. Although the ab-

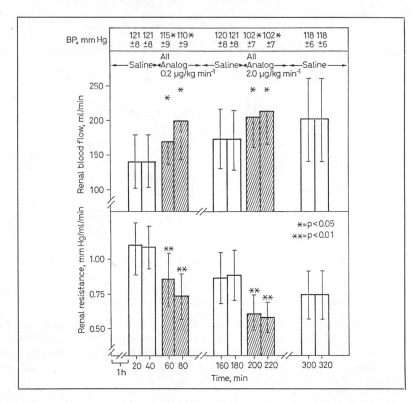

Fig. 6. Changes in mean arterial pressure (BP), renal blood flow and renal resistance in dogs with high output failure in response to administration of Sar[1]-Ala[8]-angiotensin II into the renal artery. Values are means ± SEM, n = 5. Reprinted with permission of the *American Journal of Physiology* [12].

solute level of cardiac output is high, the blood flow to the peripheral organs and tissues is inadequate to meet the needs of the body for normal functioning and a syndrome of congestive failure supervenes.

The present experiments with Sar[1]-Ala[8]-angiotensin II were undertaken to determine the role of angiotensin II in the pathogenesis of high output failure. In 1964, however, even before blocking agents for the renin-angiotensin system were available, bilateral nephrectomy was shown to produce a striking drop in arterial pressure in dogs with this type of experimental high output failure, and saline extracts of the animal's two kidneys, given intravenously, increased arterial pressure to, or nearly to,

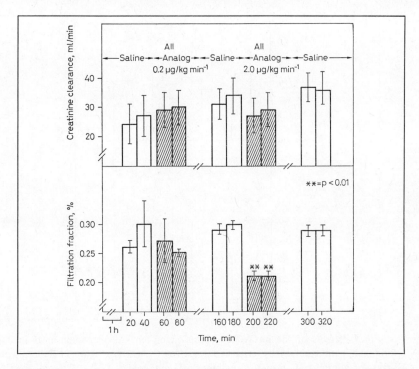

Fig. 7. Changes in renal Na excretion (E_{Na}), creatinine clearance and filtration fraction in dogs with high output failure before and after infusion of Sar[1]-Ala[8]-angiotensin II into the renal artery. Values are means ± SEM, n = 5. Reprinted with permission of the *American Journal of Physiology* [12].

the control level [13]. In a more sophisticated experiment, Sar[1]-Ala[8]-angiotensin II was given to conscious dogs with high output failure [12]. A striking drop in arterial pressure occurred during infusion of the angiotensin II analogue and blood pressure returned to the control level during the recovery period (fig. 5). A threefold increase in PRA occurred from the high control level of PRA which was secondary to the fistula. These findings show the importance of angiotensin II in maintaining arterial pressure in high output failure.

Intrarenal arterial infusion of Sar[1]-Ala[8]-angiotensin II revealed that angiotensin II contributes to the decrease in renal blood flow in high output failure as it does in low output cardiac failure [12]. With infusion of the angiotensin II analogue, an increase in renal blood flow occurred

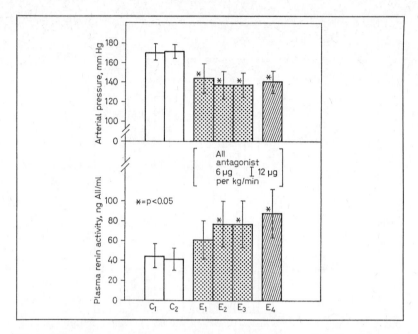

Fig. 8. Effects of the intravenous infusion of Sar¹-Ala⁸-angiotensin II in conscious malignant 1-kidney hypertensive dogs: for C_1, E_1, see figure 2. Experimental E_4 was for administration of the angiotensin II antagonist at 12 μg/kg/min. Reprinted with permission of the Society of Experimental Biology and Medicine [14].

while renal resistance fell (fig. 6); the response occurred even with the very low dose of 0.2 μg/kg/min of the blocking agent. These observations show that the kidney plays a prominent role in the maintenance of peripheral resistance and arterial pressure in high output failure.

While an increase in renal blood flow occurred with angiotensin II blockade, glomerular filtration rate (C_{cr}) was unchanged so filtration fraction (FF) fell (fig. 7) [12]. Also, the very low rate of renal sodium excretion was unchanged by angiotensin II blockade, a finding which suggests that angiotensin II has no direct influence on renal tubular reabsorption of sodium.

The role of the renin-angiotensin system in the pathogenesis of both one- and two-kidney hypertension in the dog has been studied by angiotensin II blockade. Intravenous infusion of Sar¹-Ala⁸-angiotensin II into conscious animals with so-called one-kidney malignant hypertension produced a fall in arterial pressure and an increase in PRA (fig. 8) [14]. In

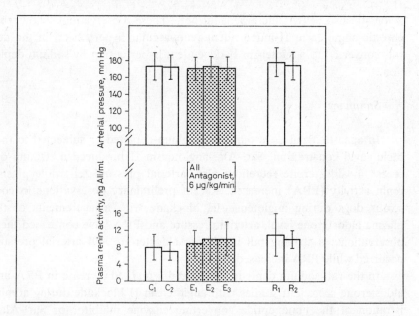

Fig. 9. Effects of the intravenous infusion of Sar[1]-Ala[8]-angiotensin II in conscious chronic 1-kidney hypertensive dogs; for C_1, E_1, see figure 2. Reprinted with permission of the Society of Experimental Biology and Medicine [14].

contrast, in chronic one-kidney hypertensive dogs, PRA was normal and the intravenous infusion of the angiotensin II analogue failed to influence either arterial pressure or PRA (fig. 9) [14].

Recently, the role of the renin-angiotensin system in both the acute and chronic phases of two-kidney hypertension in the dog has been studied [15]. On the 4th day of the acute phase, PRA was slightly elevated and a response to Sar[1]-Ala[8]-angiotensin II occurred; arterial pressure fell from 148 to 115 mm Hg and PRA increased from 7 to 11 ng angiotensin II/ml plasma. These results implicate the renin-angiotensin system in a causal role during the acute phase of hypertension. By use of the converting enzyme inhibitor, MILLER *et al.* [16] have also demonstrated the necessity of angiotensin II in acute one kidney renal hypertension in the dog.

In the chronic phase of two-kidney hypertension, PRA was normal and Sar[1]-Ala[8]-angiotensin II failed to decrease arterial pressure. However, sodium depletion of these animals increased PRA and a striking decrease in arterial pressure occurred with infusion of the angiotensin II antagonist; also, PRA increased further. These observations show that chronic angiotensin II-independent renovascular hypertension in the dog was converted to angiotensin II-dependent hypertension by sodium depletion.

Summary

In anaesthetized dogs that were sodium-depleted or subjected to thoracic caval constriction, Sar^1-Ala^8-angiotensin II produced a striking decrease in aldosterone secretion; also, arterial pressure fell while plasma renin activity (PRA) increased. Recent preliminary observations in conscious dogs during angiotensin II blockade with measurements of the plasma aldosterone level, arterial pressure and PRA have confirmed these observations; a striking fall in plasma aldosterone and arterial pressure occurred while PRA increased.

In the rat, sodium depletion produced a marked increase in PRA and aldosterone secretion; studies with angiotensin II blockade during administration of the nonapeptide converting enzyme inhibitor or Sar^1-Ala^8-angiotensin II demonstrated an important role for angiotensin II in mediating the increase in aldosterone secretion during sodium depletion in the rat.

In experimental high output failure secondary to a large aortic-caval fistula, angiotensin II blockade revealed that angiotensin II decreases renal blood flow and helps to maintain the level of arterial pressure; thus, the kidney participates in the compensatory action of angiotensin II to increase total peripheral resistance. Angiotensin II blockade in both one and two-kidney renal hypertensive dogs revealed that angiotensin II was important in the pathogenesis of the acute phase, but in chronic renal hypertension the mechanisms appeared to be angiotensin II-independent.

References

1 DAVIS, J. O.: Regulation of aldosterone secretion; in Handbook of physiology in endocrinology, vol. 6, chap. 7, pp. 77–106 (Williams & Wilkins, Baltimore 1975).
2 BLAIR-WEST, J. R.; COGHLAN, J. P.; DENTON, D. A.; FUNDER, J. W., and SCOGGINS, B. A.: The role of the renin-angiotensin system in control of aldosterone secretion; in TATIANA ASSAYKEEN Control of renin secretion, pp. 167–187 (Plenum Publishing, New York 1972).

3   Davis, J. O.; Ayers, C. R., and Carpenter, C. C. J.: Renal origin of an aldosterone-stimulating hormone in dogs with thoracic caval constriction and in sodium-depleted dogs. J. clin. Invest. *40:* 1466–1474 (1961).
4   Johnson, J. A. and Davis, J. O.: Effects of a specific competitive antagonist of angiotensin II on arterial pressure and adrenal steroid secretion in dogs. Circulation Res. *32–33:* suppl. I, pp. 159–168 (1973).
5   Davis, J. O.: The use of blocking agents to define the functions of the renin-angiotensin system. Clin. Sci. mol. Med. *48:* 3s–14s (1975).
6   Spielman, W. S. and Davis, J. O.: The renin-angiotensin system and aldosterone secretion during sodium depletion in the rat. Circulation Res. *35:* 615–624 (1974).
7   Davis, J. O. and Freeman, R. H.: The use of angiotensin II blockade to study adrenal steroid secretion. Fed. Proc. (in press 1976).
8   Spielman, W. S.; Davis, J. O., and Freeman, R. H.: Des-asp-1-angiotensin II; possible role in mediating the renin-angiotensin responses in the rat. Proc. Soc. exp. Biol. Med. *151:* 177–182 (1976).
9   Coleman, T. G.; McCaa, R. E., and McCaa, C. S.: Effect of angiotensin II on aldosterone secretion in the conscious rat. J. Endocr. *60:* 421–427 (1974).
10 Campbell, W. B. and Pettinger, W. H.: Control of aldosterone release by two angiotensins in the rat. Fed. Proc. *33:* 547 (1974).
11 Chiu, A. T. and Peach, M. H.: Inhibition of induced aldosterone biosynthesis with a specific antagonist of angiotensin II. Proc. natn. Acad. Sci. USA *71:* 341–344 (1974).
12 Freeman, R. H.; Davis, J. O.; Spielman, W. S., and Lohmeier, T. E.: High-output heart failure in the dog: systemic and intrarenal role of angiotensin II. Am. J. Physiol. *229:* 474–478 (1975).
13 Davis, J. O.; Urquhart, J.; Higgins, J. T., jr.; Rubin, E. C., and Hartroft, P. M.: Hypersecretion of aldosterone in dogs with a chronic aortic-caval fistula and high output heart failure. Circulation Res. *14:* 471–485 (1964).
14 Johnson, J. A.; Davis, J. O.; Spielman, W. S., and Freeman, R. H.: The role of the renin-angiotensin system in experimental renal hypertension in dogs. Proc. Soc. exp. Biol. Med. *147:* 387–391 (1974).
15 Watkins, B. E. and Davis, J. O.: Studies of two kidney renovascular hypertension in the dog. Physiologist *18:* 438 (1975).
16 Miller, D. E., jr.; Samuels, A. L.; Haber, E., and Barger, A. C.: Inhibition of angiotensin conversion in experimental renovascular hypertension. Science *177:* 1108–1109 (1972).

Dr. J. O. Davis, Department of Physiology, University of Missouri School of Medicine, *Columbia, MO 65201* (USA)

Experimental and Clinical Studies with Converting Enzyme Inhibitor[1]

EDGAR HABER and A. CLIFFORD BARGER

Departments of Medicine and Physiology, Harvard Medical School, and Cardiac Unit, Massachusetts General Hospital, Boston, Mass.

Contents

Introduction	16
Renin and Sodium Balance	17
Studies in Normal Human Subjects	20
Experimental Renovascular Hypertension	22
Renin-Dependent Hypertension in Human Subjects	26
Conclusion	29
References	29

Introduction

The ability to measure renin activity and angiotensin II concentration by radioimmunoassay proved to be most helpful to the physiologist and clinical investigator in understanding the response of the renin-angiotensin system to a variety of interventions. However, it was not until specific inhibitors were available that it became possible to define precisely the role of renin in any given physiologic or pathologic situation. Inhibitors have now been described that interfere with the action of renin on its substrate, which block converting enzyme in its action on angiotensin I to produce angiotensin II, and compete in the interaction of angiotensin II with its receptor site in blood vessels or in the adrenal cortex.

---

[1] This work was supported by a USPHS Hypertension SCOR Grant HL-14150.

A series of peptides originally isolated from snake venom are effective inhibitors of angiotensin-converting enzyme [1]. They act to block the cleavage of angiotensin I and thereby prevent the formation of angiotensin II [2]. The synthetic nonapeptide GLU-TRP-PRO-ARG-PHE-GLN-ILE-PRO-PRO (CEI), based on the structure of one of the natural snake venom peptides [3], has now been used in both animal and human studies.

Renin and Sodium Balance

The intravenous infusion of converting enzyme inhibitor (CEI) has no discernible effect either on blood pressure or plasma renin activity in a normal experimental animal maintained on an adequate sodium intake [4–6]. The angiotensin II competitive inhibitor Sar^1-Ala^8-AII results in a brief pressor response in the normal animal, followed by the prompt return of blood pressure to normal levels even though infusion of the drug may continue [7, 8]. Sodium deprivation, however, uncovers a very dif-

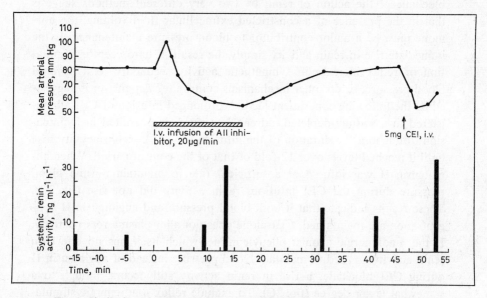

Fig. 1. Similarity of degree of hypotension induced in adrenalectomized, sodium-depleted dog given Sar^1-Ala^8-angiotensin II (20 µg/min) and the nonapeptide converting enzyme inhibitor (5 mg). Note that the initial pressor response to the angiotensin II analogue is absent with the CEI. From SAMUELS et al. [13].

ferent response [9]. Figure 1 shows the results of a typical experiment in a trained conscious dog. This animal had undergone prior adrenalectomy and was subsequently maintained on 25 mg cortisone and 1 mg DOCA. On a normal intake of sodium (50–60 mEq/day), the blocking agents did not lower blood pressure. On a 10-mEq/day diet, baseline blood pressure levels were within the normal range. The animal was alert and manifested normal activity and behaviour. Plasma renin activity was moderately elevated. Upon administration of Sar^1-Ala^8-AII, blood pressure levels first rose and then fell sharply below control levels. These haemodynamic changes were associated with a striking rise in plasma renin activity. Blood pressure soon returned to control levels coincident with a fall in renin activity. The subsequent administration of CEI resulted in a prompt fall in blood pressure without the preceding rise noted when Sar^1-Ala^8-AII was used. Renin activity also rose coincidentally with the haemodynamic change. Similar observations were made in 13 experiments in four such dogs. The results of these experiments are consistent with earlier reports [4–12].

A fall in blood pressure in a sodium-depleted animal consequent to blockade of the action of renin by two very different methods suggests that in the presence of a constricted extracellular fluid volume, the hormone must be a major contributor to blood pressure maintenance. Is the immediate rise in renin activity simply the result of baroreceptor stimulation or reflex increase in sympathetic activity secondary to a lowered blood pressure, or do other mechanisms contribute? An answer is suggested by the following experiment [13]. A prolonged infusion of CEI administered to a sodium-depleted, adrenalectomized dog resulted in hypotension throughout the duration of the infusion (fig. 2A). Renin activity rose until it reached levels over 12-fold of that of the control period. When angiotensin II was infused at a sufficient rate to maintain normal blood pressure during the CEI infusion, renin activity did not rise (fig. 2B). These results indicate that if both blood pressure and angiotensin II concentration are maintained, CEI alone does not alter plasma renin activity. If the α-adrenergic agonist phenylephrine, which itself has no effect on renin activity, is used to maintain blood pressure instead of angiotensin II during CEI blockade, a rise in renin activity still occurs, though to a somewhat lesser degree (fig. 2C). To exclude reflex β-adrenergic stimulation as the cause of renin release, a blocking dose of propranolol was given prior to the injection of the inhibitor. Even though heart rate did not change as blood pressure fell, plasma renin activity increased significant-

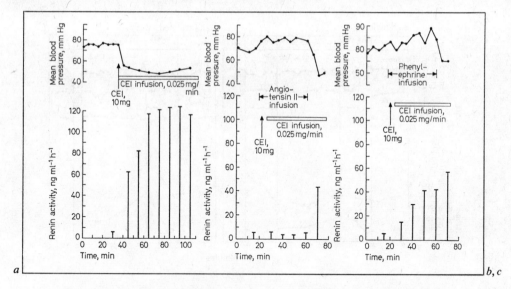

Fig. 2. A Sustained hypotension induced by infusion of nonapeptide converting enzyme inhibitor in adrenalectomized, sodium depleted animal, and subsequent rise in plasma renin activity. *B* Prevention of rise in plasma renin activity during administration of CEI in sodium depleted, adrenalectomized dog by infusion of angiotensin II sufficient to produce modest elevation of pressure. Note rapid rise of PRA following cessation of AII infusion. *C* Rise of PRA during infusion of CEI in sodium-depleted, adrenalectomized dog given an equipressor dose of phenylephrine (compare with *B*). From SAMUELS et al. [13].

ly. These experiments indicate that the rate of renin secretion is controlled by both blood pressure and angiotensin II concentration. Thus, another feedback loop is uncovered: the direct modulation of renin secretion by angiotensin II.

In earlier experiments, the intravenous infusion of angiotensin II had been shown to decrease renin release or plasma renin activity in human subjects [14] and other species [15–18]. BLAIR-WEST et al. [19] showed that this inhibition could be demonstrated at physiologic concentrations of the hormone, and that it was independent of changes in renal arterial pressure and sodium concentration. More recently, FREEDMAN et al. [20] demonstrated that infusion of either angiotensin II or the heptapeptide des-Asp[1]-angiotensin II at a rate that does not significantly affect renal blood flow, arterial blood pressure, or sodium excretion, decreases plasma renin activity. These reports further support direct negative feedback of angiotensin II on renin secretion.

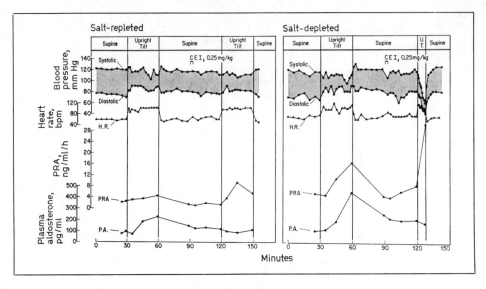

Fig. 3. Representative example of a subject (J. C.) in group I examined in the supine posture and during tilting while sodium replete and sodium depleted prior to and subsequent to the administration of converting enzyme inhibitor (CEI). HR = heart rate; PRA = plasma renin activity; PA = plasma aldosterone. From SANCHO et al. [21]. Reprinted, by permission, from *Circulation*.

Studies in Normal Human Subjects

Does the renin angiotensin system come into play only at the extremes of extracellular fluid volume depletion, or does it also have a role in blood pressure regulation at lesser degrees of physiologic stress? Studies in human subjects are most revealing [21]. Four normal young subjects were in sodium balance on a 110-mEq intake. They were first studied in the supine position, then tilted upright to 70° before and after administration of CEI. The nature and duration of the tilting stress was judiciously chosen so that none of these normal subjects would either become hypotensive or faint. During the study, heart rate and blood pressure were monitored and frequent blood samples were obtained for plasma renin activity and aldosterone concentration.

As can be seen from figure 3, prior to the administration of CEI upright tilting resulted in little haemodynamic change, a minimal narrowing of pulse pressure, and slight tachycardia. However, a rise in both plasma

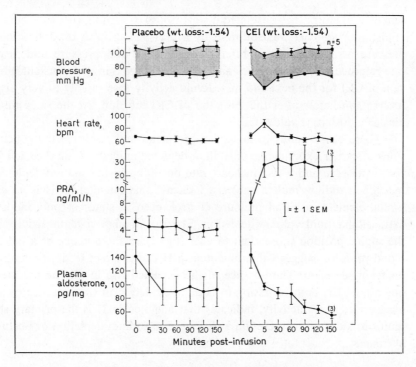

Fig. 4. Mean data of five sodium depleted subjects studied in the supine position with and without converting enzyme inhibitor. Modified from SANCHO *et al.* [21].

renin activity and plasma aldosterone concentration occurred. The administration of CEI did not result in significant haemodynamic changes either in the supine position or on tilting. Renin activity increased both in the supine position and on tilting, but no corresponding rise in aldosterone concentration was apparent. After sodium depletion, either by diet or after the administration of a diuretic, an average weight loss of 2.6 kg was observed. Supine renin and aldosterone plasma values were higher during the control period; the haemodynamic response to tilting was somewhat more marked than on the higher sodium intake. After administration of CEI, tilting was associated with a striking fall in blood pressure to hypotensive levels accompanied by an even greater elevation in renin activity; again, there was no change in plasma aldosterone concentration.

To determine the effects of prolonged infusion of CEI in the supine posture, another group of five normal subjects was given 80 mg of furosemide by mouth; they sustained a weight loss of 1.54 kg (fig. 4). The administration of CEI was followed by a transient drop in diastolic blood pressure associated with a brief tachycardia. Blood pressure and heart rate returned to control values and remained there during a constant infusion of CEI for the next 145 min. Renin activity rose rapidly to very high levels immediately after the initiation of CEI infusion. At the same time, plasma aldosterone values fell.

These observations in normal man extend and reinforce the conclusions of earlier experiments in sodium-depleted animals. A marked fall in blood pressure and a rise in heart rate on tilting in the sodium-depleted, but not in sodium-replete subjects, indicates that angiotensin II is an essential element in blood pressure control, even in states of only modest extracellular fluid volume depletion. The rise of plasma renin activity in the supine position subsequent to CEI (fig. 4), in the absence of a fall in blood pressure, suggests that angiotensin II exerts direct feedback control on renin secretion. The absence of the expected rise in plasma aldosterone after CEI, both on tilting or sodium depletion in the presence of an exaggerated renin activity, indicates that angiotensin II is the primary stimulus to aldosterone secretion in response to sodium depletion or postural change.

Experimental Renovascular Hypertension

The importance of renin in the genesis and maintenance of renovascular hypertension has been subject to conflicting interpretations and, at times, apparently irreconcilable experimental data. An elevation in renal vein renin has proven to be a most useful diagnostic test for surgically curable unilateral renal arterial stenosis. Yet in chronic experimental or clinical renovascular hypertension, plasma renin activity is often normal. Acute experimental studies are difficult to interpret because anaesthesia, surgical trauma, and blood loss all modify the response of the renin-angiotensin system. Renal function has been shown to be altered as long as one to two weeks after surgery [22]. To circumvent some of these problems, studies were performed on trained, conscious animals appropriately prepared and examined at least two weeks subsequent to surgery [23]. The animals had undergone prior unilateral nephrectomy. An external catheter allowed inflation of a cuff around the renal artery while catheters proximal and distal to the cuff permitted precise adjustment of the gradient

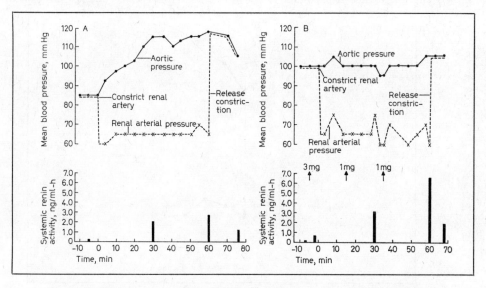

Fig. 5. Prevention of rise in systemic blood pressure after renal artery constriction by blockade of conversion of angiotensin I to angiotensin II by CEI (arrows). Data are for (A) an untreated dog and for (B) the same dog after intravenous administration of the drug. Modified from MILLER et al. [4].

created by this stenosis. Blood samples could also be obtained either in the renal vein or in the vena cava. Figure 5 summarizes the events during the first hour of constriction [4]. A substantial gradient is established between the aorta and the renal artery and is maintained constant by adjustment of the cuff. A rapid rise in mean aortic pressure to hypertensive levels occurs and is associated with an early elevation of plasma renin activity. The administration of CEI results in a prompt fall in blood pressure to near normotensive levels and a striking rise in plasma renin activity. These observations indicate that the renin-angiotensin system is responsible for the initial rise in systemic blood pressure that results from renal artery hypotension. Blockade of the conversion of angiotensin I to II promptly reduced blood pressure to normal.

During chronic renal artery constriction associated with persistent hypertension, plasma renin activity is elevated for several days, but it then returns to control levels as sodium and water retention occurs [23]. It should be of interest to determine whether or not an elevation of angi-

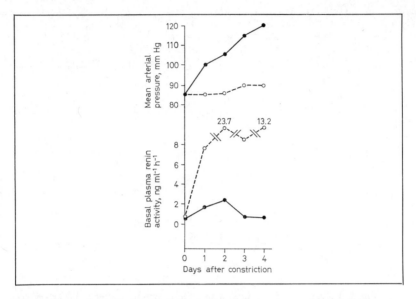

Fig. 6. Comparison of changes in mean arterial pressure and basal plasma renin activity induced by renal artery constriction in the same dog without CEI (——) and during chronic infusion of CEI (- - -). From MILLER et al. [5]. Reprinted, by permission, from the *American Journal of Physiology*.

otensin II concentration is a necessary prerequisite for chronic renovascular hypertension. This question has been addressed by initiating a constant infusion of CEI prior to renal artery constriction and then maintaining it for several days [5]. Figure 6 shows two successive experiments in the same dog. In the first series of observations, the renal artery was constricted and blood pressure rose over the course of four days from a mean pressure of 85 to 120 mm Hg. Plasma renin activity rose, peaking on the second day and then declining to control levels by the third day. During a subsequent experimental period, the artery was constricted during an infusion of CEI. Renin activity rose to much higher levels than during the first experimental period, but no elevation of blood pressure occurred over the course of four days. Such experiments provide strong evidence that elevated angiotensin II concentrations are essential in the initiation of chronic renovascular hypertension. They do not delineate the role of angiotensin II in the maintenance of established renovascular hypertension at a time when plasma renin activity is normal [23].

WAKERLIN et al. [24] reported the lowering of blood pressure in chronic renovascular hypertensive dogs as a result of immunization with renin. More recent evidence, however, has not supported the view that angiotensin II plays a significant role in the maintenance of hypertension in the chronic one-kidney Goldblatt rabbit [25] or rat [10, 26, 27]. After the initiation of chronic renovascular hypertension in the dog, a single intravenous injection of CEI produces a significant fall in blood pressure during the first three days. After the fourth day, the effect is progressively less marked [5]. These results are in essential agreement with those of BUMPUS et al. [9] who employed the Sar^1-Ile^8 analogue of angiotensin II. They reported that blockade was effective in reducing blood pressure in the conscious dog within three to six days after renal artery constriction but not later. Similarly, JOHNSON et al. [11] reported that Sar^1-Ala^8-AII did not lower blood pressure in one-kidney Goldblatt dogs two to seven weeks after constriction of the renal artery. All of these observations further confirm the earlier report of PALS et al. [10] that the blockade of angiotensin II in the one-kidney Goldblatt rat lowers blood pressure only within the first two weeks after an artery constriction. Although BING and NIELSON [12] suggested that the Sar^1-Ala^8-AII analogue was effective in lowering blood pressure one to two months after renal artery constriction in the rat, three of their seven animals did not have a drop in blood pressure, and in only one animal did blood pressure fall below 160 mm Hg at the time of the maximum effect of the drug. Thus, the decreasing effectiveness of CEI and angiotensin II antagonists in lowering blood pressure in chronic one-kidney renovascular hypertension suggests that other factors play an increasingly important role with time in the maintenance of elevated blood pressure. TOBIAN et al. [28] and GUYTON et al. [29] have stressed the importance of sodium and water retention in the maintenance of the elevated blood pressure of chronic renovascular hypertension. TAGAWA et al. [23] have shown that in a unilaterally nephrectomized dog with a normal sodium intake, sodium retention and increased water intake followed the chronic constriction of the renal artery. GAVRAS et al. [8] showed that in sodium-depleted one-kidney Goldblatt rat, Sar^1-Ala^8-AII resulted in a large drop in blood pressure, whereas after sodium repletion no lowering in blood pressure could be demonstrated with this agent.

The sum of these observations indicates that there are two phases of renovascular hypertension: Initially elevated blood pressure is maintained by the direct pressor actions of angiotensin II; during a later chronic phase, blood pressure is maintained largely as a result of hypervolaemia, mediated by sodium and water retention. Whether sodium retention is simply the direct effect of diminished renal arterial pressure on kidney function or is mediated indirectly via angiotensin II through its action on mineralocorticoid secretion remains to be determined. The lack of response to single injections of either CEI or angiotensin II analogs in chronic renovascular hypertension is not decisive. Even upon release of renal artery constriction, a period of 24 h is required to effect a diuresis with a consequent normalization of blood pressure [23]. A brief removal of angiotensin II effect would seem insufficient to perturb this long-term homeostatic adjustment. A critical experiment, which has not yet been reported, is the chronic infusion of either a receptor blocker (angiotensin II analogue) or CEI after the establishment of chronic renovascular hypertension. If angiotensin II is ultimately responsible for maintenance of hypertension, a sodium diuresis should occur with a return of blood pressure to normal levels. Thus, the primary role of the renal angiotensin system in the maintenance of chronic renovascular hypertension has not yet been determined.

Renin-Dependent Hypertension in Human Subjects

While the aetiologic role of the renin-angiotensin system in chronic renovascular hypertension remains to be established, it is generally agreed that the determination of renin activity in renal venous blood is a most useful diagnostic aid. When renal vein renin activity from the affected kidney is greater than that of the uninvolved side, the probability of improvement in blood pressure after surgery is considerable [30–34]. Prior stimulation of renin release has been shown to enhance selectively renin release from the affected kidney [30, 35]. As discussed previously, either a decrease in circulatory angiotensin II concentration or blockade of the angiotensin receptor stimulates renin release. Advantage can be taken of this feedback mechanism to enhance renin secretion of kidneys affected by renal artery stenosis thereby improving diagnostic accuracy. Figure 7 shows a study in a patient later demonstrated to have severe right renal artery stenosis. At the time of bilateral renal venous catheterization, a single intravenous dose of CEI was administered. An unanticipated and striking fall in blood pressure occurred. Right renal vein renin activity rose to very high levels. The small increase in left renal vein renin simply reflected the increase in systemic renin activity. The ratio between right

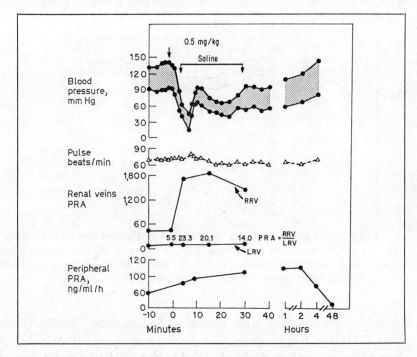

Fig. 7. Changes in systolic and diastolic blood pressure, pulse, renin activity in the renal veins, and a peripheral vein and plasma aldosterone concentration in a 54-year-old man with chronic hypertension in response to converting enzyme inhibitor (CEI). Blood pressure during the previous year had averaged 170/100 mm Hg despite therapy with hydrochlorthiazide, triamterene, hydralazine, and guanethidine. Furosemide (80 mg) had been administered 24 h prior to the study. Arrow indicates the time of administration of CEI. From the unpublished data of DUHME *et al.*, modified from *Lancet i:* 408 (1974).

and left renal venous renin activity, which is utilized as a diagnostic index, rose from 5.5 to 14.0. While the diagnosis of unilateral renal artery stenosis was not difficult to make in this patient, a marked selective stimulation of this kind may be of value in many patients who exhibit marginal lateralization in renal venous renin activities. The undesirable hypotension experienced by this patient was probably the result of sodium depletion by diuretics prior to CEI administration. Hypotension has not occurred in individuals in normal sodium balance.

Several investigators have suggested that a fall in blood pressure in response either to CEI or angiotensin II analogues may be of value in

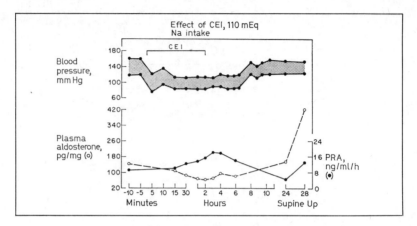

Fig. 8. Blood pressure, plasma renin activity (PRA) and plasma aldosterone in response to converting enzyme inhibitor (CEI) in a 28-year-old woman with a history of chronic bilateral pyelonephritis who was studied while in balance on 110 mEq Na intake daily. From the unpublished data of SANCHO *et al.*

identifying patients with renin-dependent hypertension [36–39]. GAVRAS *et al.* [39] stress that sensitivity of detection may increase if subjects have been previously sodium-depleted. A fall in blood pressure under these circumstances must be interpreted with the greatest caution. What is revealed is simply that, under the circumstances of testing, blood pressure maintenance is in part dependent on angiotensin II. This may be the normal physiologic state in volume depletion. If a sodium-depleted normal human subject or animal becomes hypotensive when treated with a converting enzyme or angiotensin blocker, there is no reason not to assume that patients in whom the aetiology of hypertension is independent of renin may not respond in the same way. Utilization of inhibitors in the identification of renin-dependent hypertension may require that testing be carried out while subjects are in normal sodium balance. An example is shown in figure 8. The patient is a 28-year-old woman with chronic bilateral pyelonephritis and persistent hypertension. She was studied while in sodium balance on a daily 110-mEq sodium diet. Plasma renin activities were abnormally elevated (normal values on this diet are 1.02 ± 0.21 ng/ml/h). The administration of CEI resulted in a prompt fall in blood pressure to the normotensive range. There was a further rise in plasma renin activity, accompanied by a fall in plasma aldosterone that confirmed blockade of angiotensin production.

Conclusion

The advent of inhibitors of the renin-angiotensin-aldosterone system has now permitted its place in cardiovascular homeostasis to be determined accurately. While angiotensin II does not seem to play an irreplaceable role in the normal animal or human subject in a normal state of sodium balance, it becomes a central element in the maintenance of blood pressure after sodium deprivation. Angiotensin II is essential in the initiation of renovascular hypertension. When free sodium and fluid intake is permitted in the unilaterally nephrectomized Goldblatt hypertensive animal, angiotensin II loses its importance in the maintenance of chronic hypertension. In a similar fashion, compensatory fluid retention in experimental congestive failure requires the mediation of angiotensin II. In its absence, haemodynamic exquilibrium cannot be restored and blood pressure remains at hypotensive levels.

Angiotensin antagonists may be most useful clinically in identifying renin-dependent hypertensives, though great caution must be exercised in interpreting a blood pressure fall in a sodium-depleted subject, since this is also seen in normal individuals. These compounds may be particularly helpful in the diagnosis of unilateral renal artery stenosis, since renin secretion on the affected side is selectively stimulated. Studies with antagonists have also confirmed that angiotensin II is the major stimulus to aldosterone secretion in response to posture or sodium deprivation and that angiotensin II acts as a feedback repressor of renin.

References

1. FERRERIA, S. H.: A bradykinin-potentiating factor (BPF) present in the venom of *Bothrops jararaca*. Br. J. Pharmac. Chemother. *24:* 163–169 (1965).
2. BAKHLE, Y. S.: Inhibition of angiotensin I converting enzyme by venom peptides. Br. J. Pharmacol. *43:* 252–254 (1971).
3. ONDETTI, M. A.; WILLIAMS, N. J.; SABO, E. F.; PLUSCEC, J.; WEAVER, E. R., and KOCY, O.: Angiotensin-converting enzyme inhibitors from the venom of *Bothrops jararaca:* isolating, elucidation of structure, and synthesis. Biochemistry, N. Y. *10:* 4033–4039 (1971).
4. MILLER, E. D., jr.; SAMUELS, A. I.; HABER, E., and BARGER, A. C.: Inhibition of angiotensin conversion in experimental renovascular hypertension. Science *177:* 1108–1109 (1972).
5. MILLER, E. D., jr.; SAMUELS, A. I.; HABER, E., and BARGER, A. C.: Inhibition of

angiotensin conversion and prevention of renovascular hypertension. Am. J. Physiol. *228:* 448–453 (1975).

6   BIANCHI, A.; EVANS, D. B.; COBB, M.; PESCHKA, M. T.; SCHAEFFER, T. R., and LAFFAN, R. J.: Inhibition by SQ 20881 of vasopressor response to angiotensin I in conscious animals. Eur. J. Pharmacol. *23:* 90–96 (1973).

7   JOHNSON, J. A. and DAVIS, J. O.: Effects of a specific competitive antagonist of angiotensin II on arterial pressure and adrenal steroid secretion in dogs. Circulation Res. *32:* suppl. 1, pp. 159–168 (1973).

8   GAVRAS, H.; BRUNNER, H. R.; VAUGHAN, E. D., jr., and LARAGH, J. H.: Angiotensin-sodium interaction in blood pressure maintenance of renal hypertensive and normotensive rats. Science *180:* 1369–1371 (1973).

9   BUMPUS, F. M.; SEN, S.; SMEBY, R. R.; SWEET, C. S.; FERRARIO, C. M., and KHOSLA, M. C.: Use of angiotensin II antagonists in experimental hypertension. Circulation Res. *32:* suppl. 1, pp. 150–158 (1973).

10 PALS, D. T.; MASUCCI, F. D.; DENNING, G. S., jr.; SIPOS, F., and FESSLER, D. C.: Role of the pressor action of angiotensin II in experimental hypertension. Circulation Res. *29:* 673–681 (1971).

11 JOHNSON, J. A.; DAVIS, J. O.; SPIELMAN, W. S., and FREEMAN, R. H.: The role of the renin-angiotensin system in experimental renal hypertension in dogs. Proc. Soc. exp. Biol. Med. *147:* 387–391 (1974).

12 BING, J. and NIELSON, K.: Role of the renin-system in normo- and hypertension. Effect of angiotensin inhibitor (1-sar-8-ala-angiotensin II) on the blood pressure of conscious or anesthetized normal, nephrectomized and renal hypertensive rats. Acta path. microbiol. scand. A *81:* 254–262 (1973).

13 SAMUELS, A. I.; MILLER, E. D., jr.; FRAY, J. C. S.; HABER, E., and BARGER, A. C.: Renin-angiotensin antagonists and the regulation of blood pressure. Fed. Proc. Fed. Am. Socs exp. Biol. (in press).

14 CHAMPLIN, J. DE; GENEST, J.; VEYRATT, R., and BOUCHER, R.: Factors controlling renin in man. Archs intern. Med. *117:* 355–363 (1966).

15 VANDER, A. J. and GEELHOED, G. W.: Inhibition of renin secretion by angiotensin II. Proc. Soc. exp. Biol. Med. *120:* 399–403 (1965).

16 BUNAG, R. D.; PAGE, I. H., and McCUBBIN, J. W.: Inhibition of renin release by vasopressin and angiotensin. Cardiovasc. Res. *1:* 67–73 (1967).

17 TANAKA, K.; OMAE, T.; HATTORI, N., and KATSUKI, S.: Renin release from ischemic kidneys following angiotensin infusion in dogs. Jap. Circulation J. *33:* 235–241 (1969).

18 SHADE, R. E.; DAVIS, J. O.; JOHNSON, J. A.; GOTSHALL, R. W., and SPIELMAN, W. S.: Mechanism of action of angiotensin II and antidiuretic hormone on renin secretion. Am. J. Physiol. *224:* 926–929 (1973).

19 BLAIR-WEST, J. R.; COGHLAN, J. P.; DENTON, D. A.; FUNDER, J. W.; SCOGGINS, B. A., and WRIGHT, R. D.: Inhibition of renin secretion by systemic and intrarenal angiotensin infusion. Am. J. Physiol. *220:* 1309–1315 (1971).

20 FREEMAN, R. H.; DAVIS, J. O., and LOHMEIER, T. E.: Des-1-Asp-angiotensin II: possible intrarenal role in homeostasis in the dog. Circulation Res. *37:* 30–34 (1975).

21 SANCHO, J.; RE, R.; BURTON, J.; BARGER, A. C., and HABER, E.: The role of the

renin-angiotensin-aldosterone system in cardiovascular homeostasis in normal human subjects. Circulation *53:* 400–405 (1976).

22 THORBURN, G. D.; KOPALD, H. H.; HERD, J. A.; HOLLENBERG, M.; O'MORCHOE, C. C. C., and BARGER, A. C.: Intrarenal distribution of nutrient blood flow determined with krypton[85] in the unanestheized dog. Circulation Res. *13:* 290–307 (1963).

23 TAGAWA, H.; GUTMANN, F. D.; HABER, E.; MILLER, E., jr.; SAMUELS, A. I., and BARGER, A. C.: Reversible renovascular hypertension and renal arterial pressure. Proc. Soc. exp. Biol. Med. *146:* 975–982 (1974).

24 WAKERLIN, G. E.; BRID, R. B.; BRENNAN, B. B.; FRANK, M. H.; KREMEN, S.; KUPERMAN, I., and SKOM, J. H.: Treatment and prophylaxis of experimental renal hypertension with 'renin'. J. Lab. clin. Med. *41:* 708–728 (1953).

25 ROMERO, J. C.; HOOBLER, S. W.; KOZAK, T. J., and WARZYNSKI, R. J.: Effect of antirenin on blood pressure of rabbits with experimental renal hypertension. Am. J. Physiol. *225:* 810–817 (1973).

26 BRUNNER, H. R.; KIRSHMAN, J. D.; SEALEY, J. E., and LARAGH, J. H.: Hypertension of renal origin: evidence for two different mechanisms. Science *174:* 1344–1346 (1971).

27 KRIEGER, E. M.; SALGADO, H. C.; ASSAN, C. J.; GREENE, L. L. J., and FERREIRA, S. H.: Potential screening test for detection of overactivity of renin-angiotensin system. Lancet *i:* 269–271 (1971).

28 TOBIAN, L.; COFFEE, K., and MCCREA, P.: Contrasting total exchangeable sodium levels in rats with two different types of Goldblatt hypertension. J. Lab. clin. Med. *66:* 1027–1028 (1965).

29 GUYTON, A. C.; COLEMAN, T. G., and GRANGER, H. J.: Circulation: overall regulation. A. Rev. Physiol. *34:* 13–46 (1972).

30 MICHELAKIS, A. M.; FOSTER, J. H.; LIDDLE, G. W.; RHAMY, R. K.; KUCHEL, O., and GORDON, R. D.: Measurement of renin in both renal veins: its use in diagnosis of renovascular hypertension. Archs intern. Med. *120:* 444–448 (1967).

31 WINER, B. M.; LUBBE, W. F.; SIMON, M., and WILLIAMS, J. A.: Renin in the diagnosis of renovascular hypertension. J. Am. med. Ass. *202:* 121–128 (1967).

32 FITZ, A.: Renal venous determinations in the diagnosis of surgically correctable hypertension. Circulation *36:* 942–950 (1967).

33 AMSTERDAM, E. A.; COUCH, N. P.; CHRISTLIEB, A. R.; HARRISON, J. H.; CRANE, C.; DOBRZINSKY, S. J., and HICKLER, R. B.: Renal vein renin activity in the prognosis of surgery for renovascular hypertension. Am. J. Med. *47:* 860–868 (1969).

34 LARAGH, J. H.; SEALY, J. E.; BUHLER, F. R.; VAUGHAN, E. D.; BRUNNER, H. R.; GAVRAS, H., and BAER, L.: The renin axis and vasoconstrictor volume analysis for understanding and treating renovascular and renal hypertension. Am. J. Med. *58:* 4–13 (1975).

35 STRONG, C. G.; HUNT, J. C.; SHEPS, S. G.; TUCKER, R. M., and BERNATZ, P. E.: Renal venous renin activity: enhancement of sensitivity of lateralization by sodium depletion. Am. J. Cardiol. *27:* 602–611 (1971).

36 BRUNNER, H. R.; GAVRAS, H.; LARAGH, J. H., and KEENAN, R.: Angiotensin II

blockade in man by Sar[1]-Ala[8]-angiotensin II for understanding and treatment of high blood pressure. Lancet *ii:* 1045–1048 (1973).
37 DONKER, A. J. M. and LEENEN, F. H. H.: Infusion of angiotensin II analogue in two patients with unilateral renovascular hypertension. Lancet *ii:* 1535–1537 (1974).
38 STREETEN, D. H. P.; ANDERSON, G. H.; FREIBERG, J. M., and DALAKOS, T. G.: Use of an angiotensin II antagonist (saralasin) in the recognition of 'angiotensinogenic' hypertension. New Engl. J. Med. *292:* 657–662 (1975).
39 GAVRAS, H.; BRUNNER, H. R.; LARAGH, J. H.; SEALEY, J. E.; GAVRAS, I., and VUKOVICH, R. A.: An angiotensin converting enzyme inhibitor to identify and treat vasoconstrictor and volume factors in hypertensive patients. New Engl. J. Med. *291:* 817–821 (1974).

Dr. E. HABER, Departments of Medicine and Physiology, Harvard Medical School, Massachusetts General Hospital, *Boston Mass.* (USA)

Section II: Studies in Experimental Animals

In STOKES and EDWARDS: Drugs Affecting the Renin-Angiotensin-Aldosterone System. Use of Angiotensin Inhibitors
Prog. biochem. Pharmacol., vol. 12, pp. 33–40 (Karger, Basel 1976)

Comparative Studies of the Humoral and Arterial Pressure Responses to Sar^1-Ala^8-, Sar^1-Ile^8 and Sar^1-Thr^8-Angiotensin II in the Trained, Unanaesthetized Dog[1]

EMMANUEL L. BRAVO, MAHESH C. KHOSLA and F. MERLIN BUMPUS

Research Division, The Cleveland Clinic Foundation, Cleveland, Ohio

Contents

Introduction	33
Methods	34
Results	35
Effect on Arterial Blood Pressure	35
Effect on PRA	38
Effect on Plasma Aldosterone	38
Discussion	38
Summary	39
References	39

Introduction

The finding that replacement of the aromatic side-chain in phenylalanine in position 8 of angiotensin II (Asp-Arg-Val-Tyr-Ile-His-Pro-Phe) with an aliphatic side-chain invokes antagonistic properties [1], led to the synthesis of a number of analogues of angiotensin II as antagonists of the pressor and myotropic response of angiotensin II [2]. At least two of these antagonists, Sar^1-Ala^8- and Sar^1-Ile^8-angiotensin II, have been tested in hypertensive humans [3–5] and in animals with experimental renovascular hypertension [6, 7]. The results obtained indicate that they reduce

---

[1] Supported in part by a grant from the National Heart, Lung and Blood Institute (6835) and a grant from the American Heart Association, Northeastern Ohio Affiliate (3050R).

blood pressure in those types of hypertension with high renin activity and may find clinical use. However, these antagonists suffer from a number of limitations: (a) They elicit an initial agonist activity, i.e. pressor or aldosterone-stimulating effect [8, 9], which is equal to 1–2% of the parent hormone. Part of the pressor activity has been shown to be due to the release of catecholamines [10]; (b) they also cause increases in plasma renin activity which, it has been speculated, are due to inhibition of the negative feedback effect of angiotensin II on renin release [11]; (c) all have a short *in vivo* half-life [11]. In view of these limitations, our aim has been to find potent antagonists which should be specific (or tissue-oriented) and should be devoid of the above side-effects. Comparative infusion studies in rats with several analogues indicated that Sar^1-Thr^8-angiotensin II showed the lowest agonist to antagonist ratio [12], while perfusion studies in cat adrenals indicated that this analogue did not induce the secretion of catecholamines [13]. The present investigation is an attempt to compare this analogue with Sar^1-Ala^8- and Sar^1-Ile^8-angiotensin II in dogs for (a) its specificity of antagonistic action, (b) comparative dose-ratio for maximal antagonist effect, and (c) its effect upon arterial blood pressure, renin secretion and plasma aldosterone.

Methods

Male mongrel dogs, weighing 20–25 kg, were used for these studies. Under sterile conditions, a polyethylene catheter was inserted into the aorta through an iliac artery to allow direct recording of arterial blood pressure and blood sampling. After recovery from surgery, the dogs received either low or normal dietary sodium for 2–3 weeks. During this period of dietary adjustment, the dogs were brought twice weekly to a quiet isolated room where the experiments were to be conducted. The dogs were taught to lie down quietly while their arterial blood pressures were recorded. Within two weeks, the dogs were sufficiently conditioned to their surroundings.

For all experiments, a scalp vein needle was placed in a leg vein for infusion of peptides. Arterial blood pressure was monitored constantly on a Sanborn recorder connected to a pressure transducer. All solutions were prepared fresh in 0.9% NaCl and given at a rate of 1.0 ml/min with a constant infusion pump. At no time during the studies did the dogs receive more than 100 ml of normal saline (i.e. 15.4 mEq sodium).

Experiment 1: Comparative effects of Sar^1-Ala^8-, Sar^1-Ile^8- and Sar^1-Thr^8-angiotensin II on arterial blood pressure, plasma renin activity (PRA), and plasma aldosterone concentration (PAC) were studied in five dogs on sodium restriction for 2-3 weeks. In all five dogs, urinary Na averaged 6 mEq/24 h at the time of study. Each dog received all three analogues on separate days; each had at least a two day rest period in between studies. On the experimental day, the dogs received 8.0 mg of dexamethasone phosphate intramuscularly 1 h prior to the study. Control measurements for plasma sodium and potassium, PRA and PAC were done twice 15 min apart. Infusion of the analogue was begun at a dose of 1.0 μg/kg/min and increased at 15-min intervals to 5.0 and 10.0 μg/kg/min. Arterial blood for plasma sodium and potassium, PRA and PAC was collected 15 min after infusion of each dose level. Two recovery samples were taken at 15 and 30 min after infusion of the highest dose of analogue.

Experiment 2: Comparative agonist and antagonist activity of the three analogues on vascular smooth muscle were studied in three dogs on sodium deprivation and three dogs on normal sodium intake. Each dog received all three analogues on separate days; each had a one day rest period in between studies. The analogues were given in doses of 1.0, 5.0 and 10.0 μg/kg/min. Infusion of an analogue was begun at a dose of 1.0 μg/kg/min, continued for 15 min and then stopped. A recovery period of 30–45 min was observed before infusion of the next higher dose of analogue was begun. PRA was measured before infusion of the analogue at each dose level.

Plasma sodium and potassium were measured by flame photometry. Plasma renin activity and plasma aldosterone were measured by previously described radioimmunoassay methods [14, 15].

Statistical analysis of the data was performed using the method of paired variates [16]. Changes were considered to be significant if the p value was less than 0.05.

Results

Effect on Arterial Blood Pressure

In dogs on sodium deprivation, Sar^1-Thr^8-angiotensin II appears to be the most potent of the three analogues in reducing arterial blood pressure (table I). In doses of 1.0 μg/kg/min, Sar^1-Thr^8-angiotensin II reduced

Table I. Effects of Sar^1-Ile^8-, Sar^1-Ala^8,- and Sar^1-Thr^8-angiotensin II on mean arterial pressure, plasma renin activity and plasma aldosterone concentration in five sodium-depleted dogs

Time min	Dose of analogue μg/kg/min	Sar^1-Ile^8- (n=5)			Sar^1-Ala^8- (n=5)			Sar^1-Thr^8- (n=5)		
		MAP	PRA	PAC	MAP	PRA	PAC	MAP	PRA	PAC
−15		76±1	8.5± 1.5	21±2	78±1	7.4±0.2	23±2	79±2	7.4± 0.4	22± 2
0		77±1	7.7± 1.3	20±2	77±1	7.7±0.2	19±2	78±2	7.5± 0.9	17± 4
+15	1	85±1*	6.2± 0.5*	44±4*	76±1	8.4±0.5	42±1*	54±1*	26.8± 4.6*	24± 7
30	5	66±1*	31.3± 7.1*	40±3*	59±1*	40.0±2.3*	45±1*	52±1*	74.7± 7.3*	24± 7
45	10	55±1*	55.3±11.8*	42±5*	55±1*	41.0±3.2*	49±4*	52±1*	75.3± 7.8*	26± 8
60		73±1	24.4± 3.4*	46±6*	74±2	12.7±1.2*	50±5*	78±2*	55.2±10.3*	23± 6
75		80±1	14.3± 1.4*	45±5*	79±1	9.4±0.4*	45±2*	80±2*	23.0± 1.5*	28±10

MAP = Mean arterial pressure (mm Hg); PRA = plasma renin activity (ng/ml); PAC = plasma aldosterone concentration (ng/100 ml).
* Statistically significant from control.

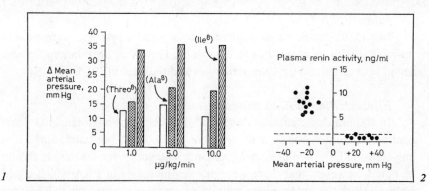

Fig. 1. Agonist activity of angiotensin II antagonists in normal sodium-repleted dogs. Each bar represents the mean value for three dogs. For Sar^1-Ile^8-angiotensin II the values are significantly higher than those for the two other antagonists. For Sar^1-Thr^8-angiotensin II the value at 10 μg/kg/min is significantly lower than those for either Sar^1-Ala^8- or Sar^1-Ile^8-angiotensin II.

Fig. 2. Relationship between basal PRA and mean arterial pressure.

mean arterial pressure by a mean of 24 mm Hg; Sar^1-Ile^8-angiotensin II maintained mean arterial pressure 8 mm Hg above control values; Sar^1-Ala^8-angiotensin II showed essentially no activity. Further, in the minimum dose which was found to be effective in reducing arterial pressure, Sar^1-Thr^8-angiotensin II was devoid of agonistic activity. Both Sar^1-Ala^8- and Sar^1-Ile^8-angiotensin II demonstrated potent initial agonistic activity (19 ± 1 and 34 ± 1 mm Hg, respectively) in the doses that were found to reduce arterial pressure subsequently. The agonistic effect was usually seen within 3 min of starting the infusion, peaked at about 4 min and was no longer evident by 8–10 min; thereafter, the depressor response became clearly demonstrated. In doses of 5.0 and 10.0 μg/kg/min, all were equally effective in reducing arterial pressure. Sar^1-Ile^8- and Sar^1-Thr^8-angiotensin II each had a much longer duration of action than Sar^1-Ala^8-angiotensin II.

In dogs on normal sodium diet, all three analogues were potent agonists on vascular smooth muscle, demonstrating essentially no antagonistic activity at all dose levels. Nevertheless, Sar^1-Thr^8-angiotensin II showed the least agonistic activity. Results of these studies are shown in figure 1.

No relationship was found between the basal values of PRA and the degree of arterial pressure reduction (fig 2). However, the analogues demonstrated antagonistic activity only when basal PRA values were elevated.

Effect on PRA

In sodium-restricted dogs, decreases in arterial pressure were usually associated with increases in PRA (table I). However, no relationship was found between blood pressure reduction and the rises in PRA.

Effect on Plasma Aldosterone

In sodium-depleted dogs, Sar^1-Ala^8- and Sar^1-Ile^8-angiotensin II demonstrated potent agonistic activity (table I). On the other hand, Sar^1-Thr^8-angiotensin II showed neither agonist nor antagonist tendencies. It should be noted, however, that there were marked increases in PRA which could have masked the antagonistic activity of these analogues on the adrenal cortex.

Discussion

Sar^1-Thr^8-angiotensin II was synthesized to study the effect of a polar group in position 8. Infusion studies in rats indicated that this analogue was equipotent with Sar^1-Ile^8-angiotensin II with the additional advantage that its initial pressor activity was 50% less than that of Sar^1-Ile^8-angiotensin II [12]. Further, its antagonistic effect lasted 2–3 h after a 30-min infusion in rats at a dose level of 250 ng/kg/min. Perfusion studies in isolated cat adrenal indicated that this peptide was devoid of catecholamine secretory activity [13].

The present studies in dogs corroborate and extend previous findings. Sar^1-Thr^8-angiotensin II was found to have no agonist effect on vascular smooth muscle in doses with significant antagonistic activity, while both Sar^1-Ala^8- and Sar^1-Ile^8-angiotensin II initially provoked significant increases in arterial blood pressure before exerting their antipressor action. Also, in marked contrast to Sar^1-Ala^8- and Sar^1-Ile^8-angiotensin II, the threonine analogue had no significant agonist activity on the adrenal cortex even when given in very large doses (i.e. at 10 µg/kg/min). It would appear, therefore, that the threonine analogue fulfills most of the criteria for a potent and specific angiotensin II antagonist and would be the antagonist most suited to study the role of the renin-angiotensin-aldosterone system in health and disease.

Sar^1-Ala^8- and Sar^1-Ile^8-angiotensin II have previously been shown to inhibit aldosterone secretion in doses similar to those used in these studies [8, 11]. The present studies, however, were performed in trained, unanaesthetized dogs which were placed on severe sodium restriction for 2–3 weeks. Additionally, the analogues were infused for shorter periods and the marked increases in PRA (and thus of angiotensin II) following their infusion could have masked their antagonistic effect on the adrenal cortex.

Although with present knowledge it is difficult to theorize on the mode of action of these antagonists or the factors which make them more specific, a recent report indicates that they inhibit the conversion of angiotensin I to angiotensin II [17, 18]. This points to the importance of investigating binding-inhibition and the inhibitory effect of these antagonists on the converting enzyme. It is interesting to note that angiotensin III has been found to be a product inhibitor of the converting enzyme [18].

Summary

The humoral and arterial blood pressure responses to Sar^1-Ala^8-, Sar^1-Ile^8- and Sar^1-Thr^8-angiotensin II were studied in sodium-depleted, trained, unanaesthetized dogs. Of the three angiotensin antagonists, Sar^1-Thr^8-angiotensin II appeared to be the best suited for clinical use. In the smallest amount that was found to be effective in reducing arterial pressure, it was devoid of agonist activity. Also, in marked contrast to Sar^1-Ala^8- and Sar^1-Ile^8-angiotensin II, Sar^1-Thr^8-angiotensin II was not shown to stimulate either catecholamine or aldosterone secretion.

References

1   KHAIRALLAH, P. A.; TOTH, A., and BUMPUS, F. M.: Analogs of angiotensin II. II. Mechanism of receptor interactions. J. med. Chem. *13:* 181–184 (1970).
2   KHOSLA, M. C.; SMEBY, R. R., and BUMPUS, F. M.: in PAGE and BUMPUS Handbook of experimental pharmacology, vol. 37, pp. 126–161 (Springer, Berlin 1974).
3   STREETEN, D. H. P.; ANDERSON, G.; FREIBERG, J. M., and DALAKOS, T. G.: Identification of angiotensinogenic hypertension in man using 1-Sar-8-Ala-angiotensin II (Saralasin, P 113). Circulation Res. *22:* suppl. I, pp. 125–132 (1975).
4   BRUNNER, H. R.; GAVRAS, H.; LARAGH, J. H., and KEENAN, R.: Angiotensin II blockade in man by Sar^1-Ala^8-angiotensin II for understanding and treatment of high blood pressure. Lancet *ii:* 1045–1066 (1973).
5   OGIHARA, T.; YAMAMOTO, T., and KUMAHARA, Y.: Clinical applications of synthetic angiotensin II analogs. Jap. Circulation J. *38:* 997–1003 (1974).

6   Bumpus, F. M.; Sen, S.; Smeby, R. R.; Sweet, C.; Ferrario, C. M., and Khosla, M. C.: Use of angiotensin II antagonists in experimental hypertension. Circulation Res. *32:* suppl. I, pp. 150–158 (1972).
7   Brunner, H. R.; Kirshman, J. D.; Sealey, J. E., and Laragh, J. H.: Hypertension of renal origin: evidence for two different mechanisms. Science *174:* 1344–1349 (1971).
8   Bravo, E. L.; Khosla, M. C., and Bumpus, F. M.: Vascular and adrenocortical responses to a specific antagonist of angiotensin II. Am. J. Physiol. *228:* 110–114 (1975).
9   Marks, L. S.; Maxwell, M. H., and Kaufman, J. T.: Saralasin bolus test. Rapid screening procedure for renin-mediated hypertension. Lancet *ii:* 784–787 (1975).
10 Munoz-Ramirez, H.; Khosla, M. C.; Bumpus, F. M., and Khairallah, P. A.: Influence of the adrenal gland on the pressor effect and antagonistic potency of angiotensin analogs. Eur. J. Pharmacol. *31:* 122–135 (1973).
11 Johnson, J. A. and Davis, J. O.: Important role of angiotensin II in the control of arterial blood pressure. Science *179:* 906–907 (1973).
12 Khosla, M. C.; Hall, M. M.; Smeby, R. R., and Bumpus, F. M.: Agonist and antagonist relationships of 1-substituted and 8-substituted analogs of angiotensin II. J. med. Chem. *17:* 1156–1160 (1974).
13 Khosla, M. C.; Munoz-Ramirez, H.; Hall, M. M.; Smeby, R. R.; Khairallah, P. A.; Bumpus, F. M., and Peach, M. J.: Synthesis of angiotensin II antagonists containing N- and O-methylated and other amino acid residues. J. med. Chem. *19:* 244–250 (1976).
14 Haber, E.; Koerner, T.; Page, L. B.; Kliman, B., and Purnode, A.: Application of a radioimmunoassay for angiotensin I to the physiologic measurements of plasma renin activity in normal human subjects. J. clin. Endocr. Metab. *29:* 1349–1355 (1969).
15 Mayes, D.; Furuyama, S.; Kem, D. C., and Nugent, C. A.: Radioimmunoassay for plasma aldosterone in men. J. clin. Endocr. Metab. *30:* 682–685 (1970).
16 Alder, H. L. and Roessler, E. B.: Introduction to probability and statistics, p. 129 (Freeman, San Francisco 1960).
17 Chiu, A. T.; Ryan, J. W.; Stewart, J. M., and Dorer, F. E.: Formation of angiotensin II by angiotensin-converting enzyme. Biochem. J. (in press).
18 Tsai, B. S.; Peach, M. J.; Khosla, M. C., and Bumpus, F. M.: Synthesis and evaluation of [Des-Asp[1]] angiotensin I as a precursor for [Des-Asp[1]] angiotensin II. J. med. Chem. *18:* 1180–1183 (1975).

Emmanuel L. Bravo, MD, Research Division, Cleveland Clinic, 9500 Euclid Avenue, *Cleveland, OH 44106* (USA)

Stimulating Effects of Angiotensin I, Angiotensin II and des-Asp^1-Angiotensin II on Steroid Production *in vitro* and its Inhibition by Sar^1-Ala^8-Angiotensin II[1]

R. Hepp, C. Grillet, A. Peytremann and M. B. Vallotton

Division of Endocrinology, Department of Medicine, Hôpital Cantonal, Geneva

Contents

Introduction	41
Methods	42
Results	42
Discussion	45
Summary	46
References	47

Introduction

In 1971, the team from the Howard Florey Institute, Melbourne, first demonstrated that the (2-8)-heptapeptide fragment of angiotensin II, des-Asp^1-AII, or as called sometimes angiotensin III, stimulated aldosterone secretion in sheep adrenals [1]. Since then, other investigators have confirmed that this heptapeptide is a potent stimulus to aldosterone biosynthesis and release, *in vivo* as well as *in vitro*. These findings stirred a great interest in the possible role of the heptapeptide as a possible physiological stimulus to steroid synthesis: it would represent the final step of the proteolytic sequence leading from renin substrate through angiotensin I (AI) and AII, before degradation to inactive fragments [2]. This prompted us to study further the steroidogenic effect of AI, AII, the (2-8)-

---

[1] Supported by the Swiss National Science Foundation (Grants No. 3.2300.74 and No. 3.7930.72) and the Swiss Foundation for Cardiology.

Table I. Preparation of isolated fasciculata cells from bovine adrenals according to the method of KLOPPENBORG *et al.* [3], as modified by SAYERS *et al.* [4]

Fasciculata slices from bovine adrenals
↓
Mechanical dispersion in Krebs-Ringer bicarbonate buffer containing 250 mg/100 ml trypsin
↓
Neutralization of trypsin with lima bean (100 mg/100 ml in KRBG)
↓
Incubation for 60 min at 37°C in an atmosphere of 95% O_2 – 5% CO_2
↓
Measurement of corticosteroids by protein-binding assay [6]
↓
Measurement of angiotensins by radioimmunoassays

---

heptapeptide and the (3-8)-hexapeptide on the fasciculata cells from bovine adrenals. These cells were chosen since they can be stimulated to synthesize corticosteroids by both ACTH and AII, which serve as control stimuli. They represent a highly sensitive preparation responding to agonist concentration in a range as low as 10^{-8} to 10^{-9} M AII or 10^{-9} to 10^{-10} M ACTH [3–5].

Methods

Table I shows the flow sheet of the experimental procedure. The glucocorticosteroids released in the medium were analysed by a competitive protein binding assay [6]. These steroids comprised approximately 60% cortisol and 30% corticosterone as judged by paper chromatography prior to assay, aldosterone accounting for less than 1%. Angiotensin I remaining in the medium was measured by radioimmunoassay and a combined value was obtained for AII plus (2-8)-heptapeptide, taking advantage of a 100% cross-reaction between these latter two compounds with the antiserum employed (fig. 1) [7, 8].

Results

A significant stimulation of corticosteroid output was noted after only 10 min of exposure of the cells to AII, the (2-8)-heptapeptide and

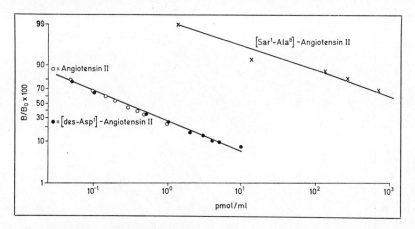

Fig. 1. Logit-log plots of the displacement curves of ^{125}I-angiotensin II by angiotensin II, des-Asp^1-AII and Sar^1-Ala^8-AII.

AI. The production of steroids tended to level off after 60 min of incubation, at which time a maximal 30-fold increase with AII and the heptapeptide and a 20-fold increase with AI had occurred at a 10^{-6} M concentration.

The stimulation of steroid production was found to be dose-dependent for all three peptides (fig. 2). The dose-response curves for AII and the heptapeptide were superimposable. They were characterized by a threshold between 10^{-9} and 10^{-8} M and a half-maximal stimulation reached at 3.5×10^{-8} M, whilst a maximal 25-fold increase in steroid production occurred at 10^{-6} M. Angiotensin I, although 3–4 times less potent than AII and the (2-8)-heptapeptide, displayed a definite effect with half-maximal stimulation occurring at 5×10^{-6} M and a 15-fold increase of basal steroid production with 10^{-6} M. By contrast, the (3-8)-hexapeptide stimulated the steroid production only slightly between 10^{-6} and 10^{-5} M.

The structural analogue and antagonist of AII, Sar^1-Ala^8-AII (saralasin or P113) was found to inhibit the steroidogenic effect of AI, AII and the (2-8)-heptapeptide. In the presence of the antagonist at a concentration of 10^{-6} M, known to block 50% of the response of the cell preparation, the dose-response curves for all three peptides were shifted in a parallel manner to the right (fig. 2). The maximal stimulation did not appear to be reduced significantly in the presence of the antagonist.

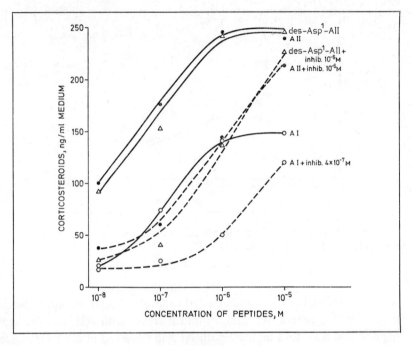

Fig. 2. Dose-response curves for angiotensins I and II and des-Asp[1]-AII on steroidogenesis from isolated fasciculata cells in the absence (solid line) and in the presence (interrupted line) of Sar[1]-Ala[8]-AII.

When the molar concentration of Sar[1]-Ala[8]-AII was modified between 10^{-9} and 10^{-5} M in the presence of each of the three peptides at a submaximal concentration of 10^{-7} M, a dose-dependent inhibition of the steroidogenic effect was observed. The half-maximal inhibition of steroidogenesis was noted at a concentration of 4×10^{-7} M with AI and 1.20 to 1.8×10^{-6} M with AII and the (2-8)-heptapeptide. As expected, Sar[1]-Ala[8]-AII did not affect the steroidogenesis induced by ACTH.

When plotted in a reciprocal manner, according to Lineweaver-Burk, the results obtained with AII, the (2-8)-heptapeptide and AI confirmed that Sar[1]-Ala[8]-AII behaved as a competitive antagonist changing the slope without affecting the V_{max} but changing the Km or A_{50} dissociation constant.

Sar[1]-Ala[8]-AII by itself possessed only a minimal agonistic activity upon fasciculata cells at 10^{-4} M, which was not statistically significant (cable II). Its effect upon aldosterone biosynthesis in the small number of

Table II. Effect of Sar^1-Ala^8-AII on steroid production by isolated bovine fasciculata cells

Mean ± SEM (n=3)	Control	Sar^1-Ala^8-AII				
		10^{-4} M	10^{-5} M	10^{-6} M	10^{-7} M	10^{-8} M
Corticosteroids, ng/ml	21.4 ±1.1	30.5 ±5.37	29.5 ±3.75	32.3 ±6.3	19.8 ±0.55	24.4 ±2.2
Aldosterone, pg/ml	285.6 ±40.0	418.0 ±35.7	338.2 ±35.0	224.3 ±18.0	227.5 ±16.2	338.6 ±35.5

glomerulosa cells remaining in the preparation of fasciculata cells was also not significant.

The possibility that AI was converted during incubation to AII or even to the (2-8)-heptapeptide had to be ruled out. The recovery of immunoreactive AI from the medium at the end of the incubation was identical whether or not cells were present. Furthermore, no generation of AII, (2-8)-heptapeptide and (3-8)-hexapeptide could be detected at the end of incubation when an antiserum fully cross-reacting with AII, (2-8)-heptapeptide and (3-8)-hexapeptide was used for the radioimmunoassay. The cross-reactivity of Sar^1-Ala^8-AII in the radioimmunoassay for AII was also checked and found to be extremely low (fig. 1). Thus, as expected [7, 8], a consequence of the replacement of phenylalanine in position 8 is that AII and Sar^1-Ala^8-AII exhibit extremely weak common immunoreactivity in both the radioimmunoassay for AII and that for Sar^1-Ala^8-AII [9].

When converting enzyme inhibitor, the nonapeptide SQ 20881 was added to the medium at concentrations ranging from 10^{-7} to 10^{-4} M; no significant alteration of steroid stimulation was observed with either AI, AII, the (2-8)-heptapeptide or ACTH.

AI and AII, when added together in the medium at submaximal doses, have additive effect on steroidogenesis, whereas at maximal doses their effect is equal to that of AII alone.

Discussion

These results demonstrate that the (2-8)-heptapeptide, known to stimulate aldosterone synthesis or to compete for binding sites in the glomerulosa cells [1, 10–13] also stimulates the steroidogenesis in fasciculata cells. The dose-response curve for AII and the heptapeptide are superimposable and similarly displaced to the right in the presence of Sar^1-Ala^8-AII. The concentration required to block 50% of the steroid production is the same for AII and the (2-8)-heptapeptide, with an inhibitor to agonist ratio of 10:1. These data demonstrating equipotency of the (2-8)-heptapeptide and AII are in agreement with those of BRECHER et al. [14] and of FREDLUND et al. [15], but not with those of CHIU and PEACH [10]. The inhibition appears competitive and is specific, ACTH stimulation persisting unchanged.

This study demonstrates also that AI is an active stimulator of steroidogenesis although 3–4 times less potent than AII and the (2-8)-heptapeptide. It possesses approximately two thirds of the intrinsic activity of these peptides and behaves as a partial agonist. Furthermore, AI appears to have intrinsic agonistic activity without having to be transformed into AII or (2-8)-heptapeptide. Just as it inhibited AII and the (2-8)-heptapeptide, Sar^1-Ala^8-AII also inhibits AI in a competitive manner, suggesting a common binding site. However, the concentration of the structural analogue required to block 50% of steroid production was 2.5 times less with a dose ratio of inhibitor to agonist of 4:1 instead of 10:1. These results are in agreement with those of BRECHER et al. [14] and of GLOSSMANN et al. [13] who found a lesser affinity of AI for binding sites in the adrenal cortex compared to AII, (2-8)-heptapeptide and Sar^1-Ala^8-AII. The difference in the response to agonist and antagonist of the vascular smooth muscle and of the adrenal cortex argues for a functional difference between the receptors in different tissues [16–18]. Sar^1-Ala^8-AII however, contrary to the finding of WILLIAMS et al. [18] in the zona glomerulosa, does not appear to exert intrinsic agonistic activity upon the fasciculata cells. In addition to its action in the adrenal medulla where it stimulates catecholamine release [19] and in the central nervous system where it could possibly have a dipsogenic effect [20], AI thus appears to act upon the zona fasciculata without prior conversion to smaller peptides.

Summary

Two of the agents known to block the renin-angiotensin-aldosterone system, namely Sar^1-Ala^8-angiotensin II and the nonapeptide SQ 20881, have been used to clarify the role of angiotensin II (AII) and its cogeners upon the steroidogenesis in isolated fasciculata cells from bovine adrenal tissue. It could be concluded that: (1) des-Asp^1-angiotensin II is as active as AII on steroidogenesis from bovine fasciculata cells; (2) angiotensin I, although less potent, stimulates steroid production without being converted to AII or des-Asp^1-AII, and (3) Sar^1-Ala^8-AII inhibits all three peptides in a competitive manner. The presence of a common receptor for all these three peptides is suggested.

References

1 BLAIR-WEST, J. R.; COGHLAN, J. P.; DENTON, D. A.; FUNDER, J. W.; SCOGGINS, B. A., and WRIGHT, R. D.: The effect of the heptapeptide (2-8) and hexapeptide (3-8) fragments of angiotensin II on aldosterone secretion. J. clin. Endocr. Metab. 32: 575–578 (1971).

2 GOODFRIEND, T. L. and PEACH, M. J.: Angiotensin III: (Des-Aspartic acid^1)-angiotensin II. Evidence and speculation for its role as an important agonist in the renin-angiotensin system. Circulation Res. 36–37: suppl. I, pp. 38–48 (1975).

3 KLOPPENBORG, P. W. C.; ISLAND, D. P.; LIDDLE, G. W.; MICHELAKIS, A. M., and NICHOLSON, W. E.: A method of preparing adrenal cell suspensions and its applicability to the in vitro study of adrenal metabolism. Endocrinology 82: 1053–1058 (1968).

4 SAYERS, G.; SWALLOW, R. L., and GIORDANO, N. D.: An improved technique for the preparation of isolated rat adrenal cells. A sensitive, accurate and specific method for the assay of ACTH. Endocrinology 88: 1063–1068 (1971).

5 PEYTREMANN, A.; NICHOLSON, W. E.; BROWN, R. D.; LIDDLE, G. W., and HARDMAN, J. G.: Comparative effects of angiotensin and ACTH on cyclic AMP and steroidogenesis in isolated bovine adrenal cells. J. clin. Invest. 52: 835–842 (1973).

6 LECLERCQ, R.; COPINSCHI, G. et FRANCKSON, J. R. M.: Le dosage par compétition du cortisol plasmatique, modification de la méthode de Murphy. Revue fr. Etud. clin. biol. 14: 815–819 (1969).

7 VALLOTTON, M. B.: Relationship between chemical structure and antigenicity of angiotensin analogue. Immunochemistry 7: 529–542 (1970).

8 VALLOTTON, M. B.: Immunogenicity and antigenicity of angiotensin I and II; in PAGE and BUMPUS Handbook of experimental pharmacology, vol. 30, pp. 185–200 (Springer, Berlin 1974).

9 PETTINGER, W. A.; KEETON, K., and TANAKA, K.: Radioimmunoassay and pharmacokinetics of saralasin in the rat and hypertensive patients. Clin. Pharmac. Ther. 17: 146–158 (1975).

10 CHIU, A. T. and PEACH, M. J.: Inhibition of induced aldosterone biosynthesis

with a specific antagonist of angiotensin II. Proc. natn. Acad. Sci. USA *71:* 341–344 (1974).

11 LOHMEIER, T. E.; DAVIS, J. O., and FREEMAN, R. H.: DES-ASP[1]-Angiotensin II: possible role in mediating responses of the renin-angiotensin system. Proc. Soc. exp. Biol. Med. *149:* 515–518 (1975).

12 CAMPBELL, W. B.; BROOKS, S. N., and PETTINGER, W. A.: Angiotensin II- and angiotensin III-induced aldosterone release *in vivo* in the rat. Science *184:* 994–996 (1974).

13 GLOSSMANN, H.; BAUKAL, A. J., and CATT, K. J.: Properties of angiotensin II receptors in the bovine and rat adrenal cortex. J. biol. Chem. *249:* 825–834 (1974).

14 BRECHER, P. I.; PYUN, H. Y., and CHOBANIAN, A. V.: Studies on the angiotensin II receptor in the zona glomerulosa of the rat adrenal gland. Endocrinology *95:* 1026–1033 (1974).

15 FREDLUND, P.; SALTMAN, S., and CATT, K. J.: Stimulation of aldosterone production by angiotensin II peptides *in vitro*: enhanced activity of the (1-sarcosine) analogue. J. clin. Endocr. Metab. *40:* 746–749 (1975).

16 JOHNSON, J. A. and DAVIS, J. O.: Effects of a specific competitive antagonist of angiotensin II on arterial pressure and adrenal steroid secretion in dogs. Circulation Res. *32–33:* suppl. I, pp. 159–168 (1973).

17 STEELE, J. M. and LOWENSTEIN, J.: Differential effects of an angiotensin II analogue on pressor and adrenal receptors in the rabbit. Circulation Res. *35:* 592–600 (1974).

18 WILLIAMS, G. H.; McDONNELL, L. M.; RAUX, M. C., and HOLLENBERG, N. K.: Evidence for different angiotensin II receptors in rat adrenal glomerulosa and rabbit vascular smooth muscle cells. Studies with competitive antagonists. Circulation Res. *34:* 384–390 (1974).

19 PEACH, M. J.: Adrenal medullary stimulation induced by angiotensin I, angiotensin II, and analogues. Circulation Res. *28:* suppl. II, pp. 107–116 (1970).

20 BRYANT, R. W. and FALK, J. L.: Angiotensin I as a dipsogen: efficacy in brain independent of conversion to angiotensin II. Pharmac. biochem. Behav. *1:* 469–475 (1973).

Dr. RENATE HEPP, Hôpital Cantonal, *CH–1211 Geneva 4* (Switzerland)

Discussion[1]

PETTINGER: I am not impressed with evidence of noradrenaline release with saralasin. There was little increase in heart rate even though we induced a degree of hypotension in some patients. If norepinephrine or epinephrine release was a major effect of saralasin, I would certainly expect some increase in heart rate when the blood pressure goes to hypotensive levels.

Second, we have assayed plasma aldosterone at 5, 10, and 20 min of saralasin infusion in hypertensive patients, and did not get a statistically significant effect on aldosterone secretion, i.e. there was no evidence for intrinsic (angiotensin-like) activity causing induction of aldosterone release.

BRAVO: The reduction in heart rate when blood pressure comes down is a very strange phenomenon indeed. It has been suggested that this is a central nervous system effect of the drug itself.

With regard to plasma aldosterone, Dr. GORDON and Dr. HOLLENBERG have recently shown that P-113 does not stimulate aldosterone in normal man in doses that block the arterial pressure response.

MAXWELL: In extensive studies on humans, we find exactly what you do in these animals, and that is that the plasma renin activity goes up abruptly, significantly and unequivocally, but only in the patients in whom the blood pressure comes down. There has been some dispute about whether the increase in PRA is simply a response to reduction of blood pressure, i.e. is a normal physiological response, or whether there are other interpretations. We also measured aldosterone immediately and over longer periods of time; we agree with Dr. PETTINGER in that we could not see a clear-cut trend in plasma aldosterone in these patients. This may result from a combination of a blockade of steroidogenesis plus an increase in PRA, that is, these effects may nullify each other.

BRAVO: Dr. MAXWELL, your data on renin are reassuring. We conducted stud-

1 Discussion of the papers of BRAVO et al. and HEPP et al. (NB: the paper of HEPP et al. was presented by Dr. VALLOTTON).

ies in which we gave the angiotensin II antagonist and raised the renin very high. If a pressor agent like phenylephrine was added to raise the blood pressure, renin levels came down in spite of the fact that the antagonist was continued. I think that the changes in PRA produced by the antagonist are not due merely to direct effects on the juxtaglomerular apparatus, but also to baroreceptor reflexes, which may override the direct effects.

FUNDER: Dr. VALLOTTON, while commenting on your last slide, you said that angiotensin I appears to be a partial agonist, and in an earlier slide this was clear from the lower plateau that you obtained compared with AII or the heptapeptide. In addition, on your last slide you showed that the effects of AII and AI were additive, over the range of doses you used, until you got up to very high doses.

Do you have evidence that AI is a partial antagonist? If you use a higher concentration of AI than AII, can you bring the AII plateau down to that of AI?

VALLOTTON: We constructed complete dose response curves with combinations of both AI and AII, and did not see a reduction in the maximal stimulation by angiotensin II. However, we have not used a higher concentration of AI than of AII.

GANTEN: We are interested in the local enzymes of the adrenal gland. Despite high isorenin activity in the adrenal gland, there is – to our surprise – no converting enzyme activity and this is in both bovine and rat adrenals. So I think that the AI data are probably due to AI and not due to conversion of AI to AII.

VALLOTTON: Thank you for confirming that.

MULROW: To comment on two aspects of your data, Dr. VALLOTTON. Firstly, the fact that AIII and AII seem to have the same potency could be interpreted as an argument against AII having to be converted to AIII before it has activity, since one would not expect the adrenal gland to be so efficient that all of the AII would be converted to AIII, when there are so many other peptidases in the adrenal destroying angiotensin II.

Secondly, I raise a theoretical question about technique. You obviously manipulate the cells before you study them, and the sensitivity of the cells to AII is probably less than *in vivo*. Is it possible that preparation of the cells slightly alters the receptor, so that it is not so specific as it might be *in vivo*? I wonder if other people who inject things into the arteries of isolated adrenals have found that AI is as potent as you find it in your system?

VALLOTTON: It is almost impossible to refute your second point. We have no proof that there has not been any change in the receptor – it is quite possible. All we can say is that these cells are very sensitive and respond in a dose-related manner to agonists and antagonists.

Regarding the first question – the transformation of angiotensin II to angiotensin III – as I said, I think it is too early to call the (2-8)-heptapeptide 'angiotensin III'. There is no strong data to show that this heptapeptide is being actually generated *in vivo*, or is the final peptide generated, acting at the receptor site in the adrenal.

MIMRAN: I think that in patients the effect of saralasin on aldosterone changes with the rate of aldosterone secretion. It seems that in patients with very high levels of aldosterone secretion – as often seen in haemodialysis patients – the aldosterone usually comes down, while in patients with normal or sub-normal plasma aldosterone concentration saralasin always has agonistic activity. In 19 studies performed

Discussion

with saralasin, 0.5 and 2.5 µg/kg/min, there was a significant negative correlation (r = –0.69, p<0.005) between control plasma aldosterone concentration (ng%) on the abcissa and change in plasma aldosterone (ng%) induced by saralasin.

PETTINGER: Dr. VALLOTTON, to respond to your comment about the formation of angiotensin III (des-Asp-AII) in the adrenal gland: Dr. CAMPBELL in our laboratory found the enzyme aminopeptidase A to be concentrated in the adrenal cortex of rats. It responds to changes in sodium balance. Salt depletion induces a two- to threefold increase in the activity of this enzyme in the adrenal cortex. It would be nice to think of this as a mechanism for enhancing sensitivity to angiotensin in the control of aldosterone secretion. However, an inconsistency occurs when DOCA and Na^+ are given to these animals. Under this circumstance, an elevation of the activity of this aminopeptidase enzyme also occurred.

STREETEN: In connection with the effects of saralasin on aldosterone levels, I would like to say that our experience parallels that of Dr. MIMRAN rather closely. When there is a high level of renin activity and consequently of aldosterone production, saralasin tends to reduce aldosterone blood levels. On the other hand, one can show that in individuals whose renin-angiotensin system has been suppressed, e.g. by the intravenous infusion of sodium chloride, saralasin will raise plasma aldosterone concentration quite rapidly.

I would like to ask Dr. BRAVO a question in connection with his interesting observation that saralasin has an agonistic action on blood pressure in some of his dogs on the low sodium diet. We have not observed an agonistic action of saralasin in normal human subjects on a low sodium diet, but we do see this very frequently in human hypertensive patients, especially of the 'low-renin' type. I wonder what the level of sodium intake was in Dr. BRAVO's dogs and if he has any information about whether the agonistic action of saralasin could be overcome by slightly greater sodium depletion in these dogs?

BRAVO: These dogs, Dr. STREETEN, were on a low sodium diet of 6 mEq/day for two to three weeks. By the time we did the studies, the urinary sodium excretion in all these dogs was about equal to intake.

In regard to the agonist effect of the antagonist, what you say is very true. As you know, we started with 1 µg/kg/min. However, if you start with smaller doses, like 0.25 µg/kg/min, and follow this with 0.5 and 1.0 µg/kg/min, you may not see the initial agonist effect. This may, in fact, be a way of avoiding the initial agonist activity on arterial pressure.

DENTON: Dr. VALLOTTON, you may remember when we described the heptapeptide effect, JOHN COGHLAN put forward the idea that the transfer may be from nonapeptide to heptapeptide. Have you tried the nonapeptide in your preparation?

Dr. BRAVO, you mentioned that your dogs are sodium-depleted, trained and unanaesthetized. What was the reason for giving them dexamethasone? Second, can I assume from the data that we would now both agree, in relation to the threonine derivative, that it will not reduce aldosterone in the sodium-deficient state, as Dr. SCOGGINS described [BLAIR-WEST et al.: Angiotensin II analogues and aldosterone in sodium-deficient sheep. Clin. Sci. mol. Med., in press 1976]?

VALLOTTON: My response is, yes, we have tried. The nonapeptide, which we have just received from Dr. BUMPUS, gives dose-response curves superimposable upon those of angiotensin I without formation of heptapeptide.

Discussion

BRAVO: The nonapeptide has about 50% of the activity of AI when given *in vivo* in the dog, and *in vitro* it is about the same. Interestingly enough, you can use as much as 50 µg of SQ 20881 in an *in vitro* system and block aldosterone production for 30 min, but after that it continues to rise. So, SQ 20881 (converting enzyme inhibitor) has some effect.

With regard to the threonine analogue, I agree that it does not reduce aldosterone secretion in the sodium-deplete state.

Question from floor: What is the ratio of antagonist to agonist effect for the threonine analogue on aldosterone secretion?

BRAVO: There was hardly any antagonist effect and hardly any agonist effect, so it is zero.

COGHLAN: A brief comment about the question of whether heptapeptide exists in blood: certainly in the sheep, if we infuse angiotensin II, about 30% of the immunoreactivity in arterial blood is in fact heptapeptide.

Second, there are risks in sampling at 15 min, because saralasin has a relatively longer half-life and will plateau more slowly.

In this session, I think we are forgetting that the peripheral level of aldosterone is determined by the ratio of secretion rate and metabolic clearance rate. In some of these studies, very large haemodynamic changes occur. Without knowing the blood clearance rate, it is very risky to interpret the effects on aldosterone levels as meaning changes in secretion rate alone.

EDWARDS: Dr. VALLOTTON, I am a little uncertain about the steroids you are actually measuring. You just mention steroidogenesis and I think a number of the discussants are equating your results with aldosterone. Could you briefly say what steroids you are measuring in your preparation?

VALLOTTON: As I said at the beginning of my presentation, steroid production was measured in the medium by the protein-binding method, so it is glucocorticoids which are measured, mainly corticosterone and cortisol – about 60 and 30%, respectively.

Effect of Adrenal Arterial Infusion of Saralasin (P113) on Aldosterone Secretion

J. R. BLAIR-WEST, J. P. COGHLAN, D. A. DENTON, H. D. NIALL, B. A. SCOGGINS and G. W. TREGEAR

Howard Florey Institute of Experimental Physiology and Medicine, University of Melbourne, Parkville, Vict.

Contents

Introduction 53
Results and Discussion 54
Conclusions 61
Summary 61
References 61

Introduction

The availability of angiotensin antagonists has made possible a critical test of the role of the renin/angiotensin system in aldosterone regulation. There are two major propositions. One, is that circulating angiotensin II is the primary cause of high aldosterone secretion in Na deficiency, with close proportion between the prevailing blood angiotensin II concentration and the aldosterone secretion rate. The other proposal, stemming from our earlier studies, is that angiotensin, though essential for high aldosterone secretion in sodium deficiency, has a supportive or permissive role, acting in concert with other factors [1, 2].

Our experiments have shown that: (a) the aldosterone biosynthetic site stimulated by angiotensin II is not the same as the site stimulated by Na deficiency [3]; (b) aldosterone secretion increases during Na deficiency though raised renin release is prevented by infusion of angiotensin II into the renal artery [4]; this aldosterone rise was not attributable to plasma K change, and (c) in conscious, Na-deficient, nephrectomized and dexamethasone-suppressed sheep, the blood aldosterone concentration falls into the basal range; however, intravenous infusion of angiotensin I that merely restores the basal blood angiotensin II concentration is sufficient to restore the high aldosterone secretion rate of Na deficiency [2].

To test between the proposals that the blood angiotensin level may be the *major proportional cause* or a *permissive contributor* to aldosterone secretion in Na deficiency, we have infused Sar^1-Ala^8-angiotensin II (P113) into the adrenal arterial blood supply of conscious sodium-deficient sheep. The rationale is that, were angiotensin the major proportional cause, P113 infusion would cause a gradual dose-dependent suppression of aldosterone secretion. Whereas, were angiotensin permissive, the response to P113 would approach an all-or-none relation.

By infusing P113 directly into the arterial blood supply of adrenal transplants [5] we were able to: (a) estimate the local blood concentration of P113, because the adrenal blood flow is known. For example, the adrenal blood flow is about 1 litre/h and therefore P113 infusion at 10 μg/h causes a local blood concentration of 10 μg/litre, and (b) maximize P113 concentration in the adrenal arterial blood and minimize its concentration in other vascular beds. Thus, for example, we avoided large effects on cardiovascular functions.

Results and Discussion

Figure 1 shows the results of 2 experiments. P113 was infused for 2 h at 10 μg/h. Aldosterone secretion rose briefly after about 15 min and then stabilized at a higher level than control. Plasma renin concentration (PRC) rose steadily during the first hour and remained at this level for at least an hour after the infusion was stopped. This reaction shows substantial uncoupling by P113 of the mechanism by which circulating angiotensin suppresses renin release. During P113 infusion for 2 h at 100 μg/h (estimated adrenal blood concentration of about 100 μg/litre), aldosterone secretion fell briefly and then continued at near the control level. PRC rose slightly during this experiment.

The system appears to be in a steady state over the second hour of infusion and later infusions were tested for about $1^1/_2$ h. We have shown that the effect of infused angiotensin II on aldosterone secretion was reversed within 1 h of stopping the infusion [6]. Therefore, $1^1/_2$ h of P113

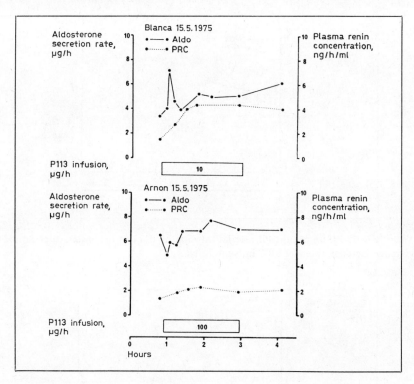

Fig. 1. Effect of adrenal arterial infusion of P113 on aldosterone secretion rate and PRC in Na-deficient sheep. Infusion rates: 10 (top panel) and 100 µg/h (bottom panel).

infusion should have been sufficient for any inhibitory effect to be observed.

P113 was infused into the adrenal artery at 1, 10, 100 and 1,000 µg/h over 5 h without any consistent effect on aldosterone secretion (fig. 2). PRC rose gradually with each step-up of infusion rate.

In a similar experiment in another animal (fig. 3), aldosterone secretion described a sawtooth pattern. Periods of lowered secretion rate followed each increase of P113 infusion rate, followed by recovery to near control level. This pattern of response was often seen in this animal. PRC rose only during infusion at 1,000 µg/h.

The results of 46 infusions of P113 in Na-deplete sheep are shown in figure 4. Results are plotted as mean aldosterone secretion rate ± SE during the control period, after 15 min and after 1–2 h at given rates of

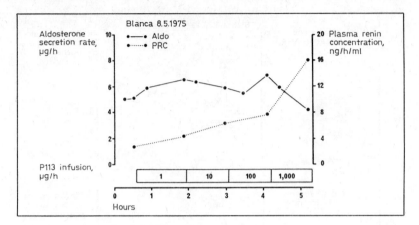

Fig. 2. Effect of adrenal arterial infusion of P113 at increasing rates on aldosterone secretion rate and PRC in Na deficiency.

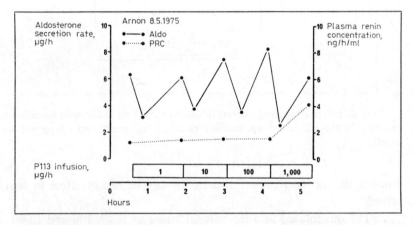

Fig. 3. Effect of adrenal arterial infusion of P113 at increasing rates on aldosterone secretion rate and PRC in Na deficiency.

P113 infusion. Infusion rate increases in decades from 10 to 1,000 ng/h at the bottom, up to 10,000–1,000,000 ng/h at the top. Throughout the range there is no evidence that aldosterone secretion was significantly inhibited.

Since adrenal blood flow was approximately 1 litre/h or less, the P113 concentrations in adrenal arterial blood can be readily calculated, e.g. 10–100 ng/h is approximately 10–100 ng/litre. The blood AII con-

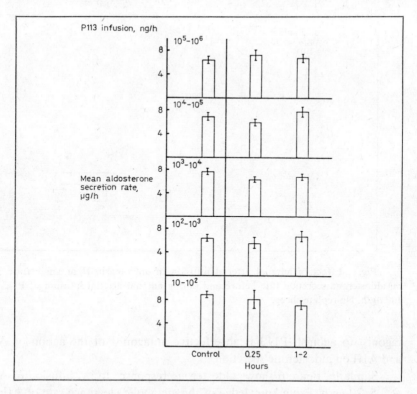

Fig. 4. Aldosterone secretion rate (mean ± SE) during control period and after 15 min and 1–2 h of adrenal arterial infusion of P113 at 10–10^6 ng/h, arranged in decades. 46 experiments in moderately Na-deficient sheep.

centration in Na-replete and mildly Na-deficient sheep falls into this range (10–100 ng/litre). In severe Na deficiency, the blood AII concentration may be 100–1000 ng/litre. Therefore, the P113 infusion rates used here caused local blood concentrations of 1,000–10,000 times the expected blood AII concentration, but aldosterone secretion was not inhibited.

Yet we have confirmed that P113 is an effective antagonist of the action of circulating angiotensin II on aldosterone secretion in the sheep. Figures 5 and 6 are representative of many observations.

Angiotensin II and III were infused at 0.5 μg/h into the adrenal arterial blood supply of Na-replete sheep (fig. 5). Each one caused a large increase of aldosterone secretion. Then P113 was infused for 2 h at 100 μg/h. In the presence of P113, the responses to AII and AIII were almost abolished. Therefore, at a concentration ratio of about 200:1, an-

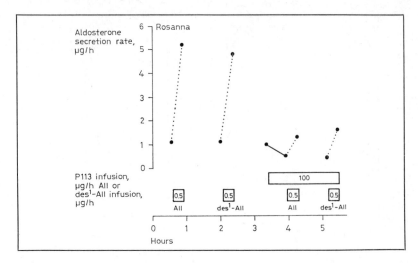

Fig. 5. Effect of adrenal arterial infusion of angiotensin II or angiotensin III on aldosterone secretion rate before and during adrenal arterial infusion of P113 at 100 µg/h. Na-replete sheep.

tagonist to agonist, P113 is an effective antagonist of the action of AII and AIII on aldosterone secretion.

Similarly, figure 6 shows aldosterone responses to i.v. infusion of AII at 2.5–20 µg/h. From knowledge of the metabolic clearance rate of AII in sheep, the expected blood AII concentrations are 25–200 ng/litre [7]. These levels are in the range observed in normal or moderately sodium-deficient sheep. Yet P113 infusions at 10 µg/h (bottom panel) or 100 µg/h (top panel) blocked aldosterone reponses to 3 levels of angiotensin II infusion in the range 2.5–20 µg/h (fig. 6). That is, a P113 concentration of 10,000 ng/litre was sufficient to inhibit the effect of a blood AII concentration of 200 ng/litre – an effective concentration ratio of about 50:1.

Since concentration ratios of 50:1 or 200:1 were sufficient to inhibit the response to exogenous AII and AIII, how are we to explain the finding that concentration ratios of 1,000–10,000:1 had no consistent effect on the high aldosterone secretion rates of Na deficiency? The possibility that aldosterone secretion in Na deficiency is entirely independent of angiotensin II seems unlikely in view of evidence that angiotensin formed by renin of renal origin is essential for the maintenance of the high aldosterone secretion rate.

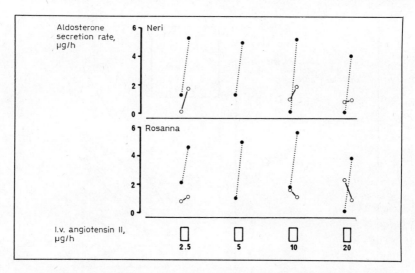

Fig. 6. Effect of i.v. infusion of angiotensin II on aldosterone secretion rate in the absence (●) and presence (O) of adrenal arterial infusion of P113 at 10 (bottom panel) and 100 µg/h (top panel). Na-replete sheep.

During P113 infusion for 1–2 h, particularly at the higher rates, PRC rose consistently though to a variable extent. This effect demonstrates the potency of P113 in inhibiting angiotensin feedback on renal receptors influencing renin release. The peripheral blood concentration of P113 must have been considerably less than the concentrations of P113 in the adrenal arterial blood. That this rate of PRC, and presumably circulating AII, could have sustained the high aldosterone secretion during P113 infusion can be discounted, because the rise doubled or perhaps trebled PRC levels and, assuming this was reflected in a two or threefold increase of blood AII concentration, would not much alter the concentration ratios of P113 to AII (of the order of 5,000:1 in the Na deficiency experiments).

The most likely explanation of the results comes from the evidence presented earlier that, in Na deficiency, aldosterone secretion does not bear a simple cause and effect proportional relationship to the prevailing blood AII concentration. However, the results could be consistent with our proposal that angiotensin may have a supportive or permissive role in concert with other factors.

At concentration ratios of about 5,000:1 (antagonist to agonist), it seems likely that even the low blood levels of AII that may provide the necessary support for high aldosterone secretion could be blocked. How-

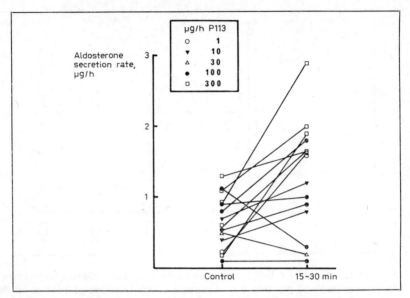

Fig. 7. Effect of adrenal arterial infusions of P113 on aldosterone secretion rate in sodium-replete sheep.

ever, figure 7 shows that P113 itself may have sufficient agonist action to provide that angiotensin-dependent support. Na-replete sheep were given adrenal arterial infusions of P113 at 1-300 µg/h. Mean aldosterone secretion rate rose from 0.7 to 1.3 µg/h over 15–30 min and the difference was significant. Cortisol secretion rate, plasma Na and PRC were unchanged during P113 infusion but plasma K fell. The result suggests that P113 may have sufficient agonist action for aldosterone secretion to support high aldosterone secretion in Na deficiency, despite virtually total blockade of circulating angiotensin.

Though we feel this latter explanation to be the most likely, it cannot be excluded that the failure of P113 to inhibit aldosterone secretion in Na deficiency might possibly be due to failure of P113 to gain access to the intra-adrenal sites of formation of angiotensin II. Our own data indicates that the renin content of whole adrenal glands of Na-replete sheep is about 4 times the renin concentration in blood, and other workers have reported higher levels. Though the physiological significance of this local renin accumulation is not known, local angiotensin II production could sustain aldosterone secretion despite effective blockade of circulating angiotensin, if its site of production were inaccessible to P113.

Conclusions

The results are against the proposal of a primary proportional causal relationship between blood AII concentration and aldosterone secretion rate in Na deficiency. The results also deny the possibility that such a primary role is played by blood-borne AIII. The results favour the proposal that high aldosterone secretion in Na deficiency may require only a low permissive blood angiotensin II concentration.

Summary

To test the role of the renin/angiotensin system in aldosterone regulation, Sar^1-Ala^8-angiotensin II (P113) was infused into the arterial blood supply of transplanted adrenal glands in conscious sheep. Effects on the aldosterone response to infused angiotensin II and III in sodium replete sheep were compared with effects in sodium deficiency.

Adrenal arterial infusion of P113 up to 1,000 µg/h for 1–2 h did not consistently alter the high aldosterone secretion rates of sodium-deficient sheep. However, infusion of P113 at 10 µg/h or more abolished aldosterone responses to angiotensin II infusion that caused high physiological blood levels of angiotensin II.

These results are against the proposal of a primary proportional causal relationship between blood angiotensin II concentration and aldosterone secretion rate in sodium deficiency. They are also against the possibility that such a primary role is played by blood-borne angiotensin III.

References

1　COGHLAN, J. P.; BLAIR-WEST, J. R.; DENTON, D. A.; SCOGGINS, B. A., and WRIGHT, R. D.: Perspectives in aldosterone and renin control. Aust. N.Z. Jl Med. 1: 178–197 (1971).
2　BLAIR-WEST, J. R.; COGHLAN, J. P.; DENTON, D. A., and SCOGGINS, B. A.: Aldosterone regulation in sodium deficiency: role of ionic factors and angiotensin II; in PAGE and BUMPUS Handbook of experimental pharmacology, pp. 337–368 (Springer, Berlin 1973).
3　BLAIR-WEST, J. R.; CAIN, M. D.; CATT, K. J.; COGHLAN, J. P.; DENTON, D. A.; FUNDER, J. W.; SCOGGINS, B. A.; STOCKIGT, J. R., and WRIGHT, R. D.: Further facets of aldosterone regulation, in JAMES and MARTINI Hormonal steroids, pp. 572–580 (Excerpta Medica, Amsterdam 1971).

4   Blair-West, J. R.; Coghlan, J. P.; Cran, E. J.; Denton, D. A.; Funder, J. W., and Scoggins, B. A.: Increased aldosterone secretion during sodium depletion with inhibition of renin release. Am. J. Physiol. *224:* 1409–1414 (1973).
5   Wright, R. D.; Blair-West, J. R.; Coghlan, J. P.; Denton, D. A.; Goding, J. R.; Nelson, J. F., and Scoggins, B. A.: The structure and function of adrenal transplant. Aust. J. exp. Biol. med. Sci. *50:* 873–892 (1972).
6   Blair-West, J. R.; Coghlan, J. P.; Denton, D. A.; Goding, J. R.; Munro, J. A.; Peterson, R. E., and Wintour, E. M.: Humoral stimulation of adrenal cortical secretion. J. clin. Invest. *41:* 1606–1627 (1962).
7   Cain, M. D.; Catt, K. J.; Coghlan, J. P., and Blair-West, J. R.: Evaluation of angiotensin II metabolism in sheep by radioimmunoassay. Endocrinology *86:* 955 (1970).

Dr. J. R. Blair-West, Howard Florey Institute of Experimental Physiology and Medicine, University of Melbourne, *Parkville, Victoria* (Australia)

In Stokes and Edwards: Drugs Affecting the Renin-Angiotensin-Aldosterone System. Use of Angiotensin Inhibitors
Prog. biochem. Pharmacol., vol. 12, pp. 63–83 (Karger, Basel 1976)

Effects of Saralasin on Renal Function in the Rat

K. G. Hofbauer, K. Bauereiss, H. Zschiedrich and F. Gross

Department of Pharmacology, University of Heidelberg, Heidelberg

Contents

Introduction	63
Materials and Methods	64
Isolated Perfused Rat Kidney	64
Acute Renal Failure	65
Substances	65
Statistical Analysis	65
Experiments	66
Isolated Perfused Rat Kidney	66
Acute Renal Failure	67
Results	68
Isolated Perfused Rat Kidney	68
Acute Renal Failur	70
Discussion	71
Intrinsic Activity of AII Antagonists	71
Renal Haemodynamics and Circulating AII	72
Intrarenal Formation of AII	73
Autoregulation and Intrarenal AII	74
Renin Release	74
Renal Prostaglandins	75
Acute Renal Failure	75
Summary	76
Acknowledgements	77
References	77

Introduction

Competitive antagonists of angiotensin II (AII) have been widely used to elucidate the role which the renin-angiotensin system (RAS) may have in the regulation of blood pressure. Considerable attention has also been paid to the effects of AII antagonists on renal haemodynamics [17, 18].

We studied the renal action of Sar^1-Ala^8-AII (P113, saralasin), a competitive antagonist of AII [66], in the isolated perfused rat kidney. Since the perfusion medium does not contain any component of the RAS, no AII reaches the kidney and an intrarenal formation of AII is unlikely, for no renin substrate is supplied. By means of such a simplified model, the intrinsic activity of saralasin as well as its antagonistic effect against infused AII can be investigated.

In another series of experiments, we evaluated the effects of saralasin in acute renal failure (ARF). It has been suggested, that the RAS participates in the pathogenesis of ARF in man and animals [11, 12], but results obtained with AII antibodies or AII antagonists were equivocal. While relatively large volumes of AII antiserum had some protective effect [70, 72], purified preparations of AII antibodies, given in a small volume, did not improve renal excretory function [48, 63].

Therefore, our experiments were initiated to elucidate the synergism between volume substitution and blockade of the AII receptors in glycerol-induced ARF. Urine volume, urine osmolality and plasma urea concentration were measured as parameters of renal excretory function during the early phase of ARF.

The discussion of the results obtained also reviews studies on the effects of AII antagonists on renal function under various conditions.

Materials and Methods

Isolated Perfused Rat Kidney

The preparation and the perfusion system have been described in detail previously [31, 32]. Male Sprague-Dawley rats weighing 180–240 g were anaesthetized with sodium pentobarbitone (Nembutal, 50 mg/kg, i.p.) and the right kidney was prepared and connected to the perfusion system without exposing it to a period of ischaemia [79].

The kidneys were perfused with a modified Krebs-Henseleit solution containing 35 g/litre of a gelatine preparation (Haemaccel, Behringwerke) at a constant pressure of 90 mm Hg in a single-pass system at 37 °C. Perfusion pressure was measured by a pressure transducer (Statham P23Db) and perfusate flow by a drop counter placed at the venous catheter.

For the determination of renin, samples of venous perfusate (50 μl) were incubated (37 °C, pH 7.2) for 30 and 60 min with 200 μl of a solution containing 500 pmol of a rat plasma renin substrate preparation [10], 3 mM 8-hydroxyquinoline, 5 mM Na_2EDTA, and 1.6 mM dimercaprol. The reaction was stopped by immediate cooling in ice water and the amount of AI formed determined by radioimmunoassay [65]. Renin concentration is expressed as the amount of AI formed during 1 h of incubation. All doses are expressed as the concentration of substances per ml of the perfusion medium.

Acute Renal Failure

The experimental procedures have been described in detail elsewhere [5]. In male Sprague-Dawley rats weighing 180–225 g, a catheter was inserted into the right jugular vein and brought to the surface at the back of the neck. During the following 4–5 days, the rats had free access to food (ssniff pellets) and demineralized water. Subsequently, they were placed into individual metabolic cages and deprived of food and water for 24 h. After the end of that period, glycerol (50% solution in demineralized water, w/w) was injected under light ether anaesthesia into the muscles of the rear limbs. During the following 8-hour period, food and water were still withheld; for the subsequent 16 h the rats had free access to water, but food was not given.

At 8 or 24 h after the injection of glycerol, the abdomen was opened by midline incision under light ether anaesthesia and blood samples were drawn from the vena cava. Urine was collected in graded tubes and the volume was determined at 2, 4, 8, and 24 h after glycerol injection. Urine osmolality was measured by an osmometer (Knauer) and plasma urea concentration was determined by the urease method (Merckotest Nr. 3334).

Substances

Asp^1-Ile^5-AII (Division of Biological Standards, Medical Research Council, London); Sar^1-Ala^8-AII (P113, saralasin) (Norwich Pharmacal Company, Norwich, N.J.).

Statistical Analysis

All values given in the text and in the figures are means ± SE. Student's *t*-test, or when appropriate, *t*-test for paired data have been used for the evaluation of significance.

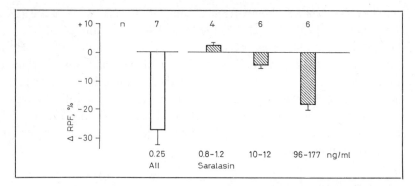

Fig. 1. Changes in renal perfusate flow (RPF) during the infusion of either AII (white bar) or various doses of saralasin (hatched bars) for 5-min periods (means ± SE).

Experiments

Isolated Perfused Rat Kidney

Experiments were performed to study the intrinsic activity of saralasin in the absence of AII, and the inhibitory action of saralasin on the renal vasoconstriction induced by infused AII. After an equilibration period of at least 50 min subsequent to the beginning of the perfusion, AII (0.1–0.3 ng/ml) was infused for a 5-min period. 10 or 15 min later saralasin was infused at various dose ranges (0.8–1.2, 10–12, and 96–177 ng/ml) for 5 min alone and for another 5 min together with AII.

Agonistic effects of saralasin were evaluated by comparing the values of renal perfusate flow (RPF) and renin release (RR) before and during the infusion of saralasin alone. The antagonistic effect of saralasin on the AII-induced renal vasoconstriction was calculated by comparing the vasoconstrictor effect of AII before and during the concomitant administration of the antagonist. (Agonistic effects of saralasin were not included in the evaluation of the AII response.)

In other experiments, the agonistic effects of high concentrations of saralasin on RPF and RR were studied in the presence of either low or high levels of AII in the perfusate. AII was infused at concentrations of 40–60 pg/ml for 5–10 min alone and for another 5 min together with saralasin. 30–40 min later, AII was infused at higher concentrations (400–1,350 pg/ml) for 5–10 min alone and for another 5 min together with saralasin (80–540 ng/ml).

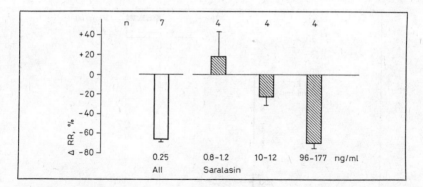

Fig. 2. Changes in renin release (RR) during the infusion of either AII (white bar) or various doses of saralasin (hatched bars) for 5-min periods (means ± SE).

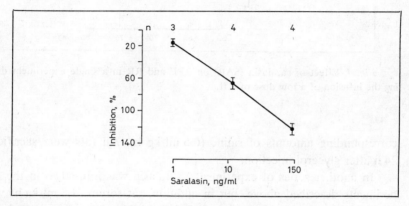

Fig. 3. Percent reduction of the AII induced vasoconstriction by various doses of saralasin (means ± SE).

Further experiments were undertaken to study if immediately after connecting the kidney to the perfusion system significant amounts of AII are still present. Saralasin was infused 5 min after the start of the perfusion (6–17 ng/ml) for a 5-min period.

Acute Renal Failure

In a first series of experiments, an infusion of saralasin (10 μg/kg min, 0.5 ml/kg h) into the jugular vein was started 30 min before the injection of glycerol and was continued for 8 h thereafter. Control rats received

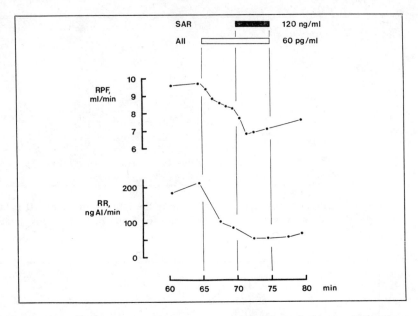

Fig. 4. Effect of saralasin (SAR) on RPF and RR in a single experiment during the infusion of a low dose of AII.

corresponding amounts of saline (0.5 ml/kg h). All rats were sacrificed 24 h after glycerol injection.

In another series of experiments, saralasin was infused as in the experiments described above, but in addition, rat serum (4.5 ml/kg h) was given during 4 h after the injection of glycerol. The respective control rats received serum only. The rats were sacrificed either 8 or 24 h after the glycerol injection.

Results

Isolated Perfused Rat Kidney

Saralasin at the two lower doses (1 and 11 ng/ml) had no significant effect on RPF, but decreased perfusate flow by 18% at the highest concentration (130 ng/ml) (fig. 1). RR was not significantly affected by the lowest dose of saralasin (1 ng/ml), but was decreased by higher doses (11 and 130 ng/ml) (fig. 2).

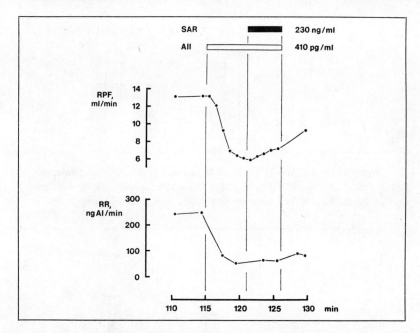

Fig. 5. Effect of SAR on RPF and RR in a single experiment during the infusion of a high dose of AII.

Saralasin inhibited the vasoconstriction induced by AII in a dose-dependent manner (fig. 3). The medium dose of saralasin, which had no effect on RPF by its own (fig. 1), diminished the AII-induced vasoconstriction by 66%. During the infusion of AII together with the highest concentrations of saralasin (130 ng/ml), a moderate vasodilation even occurred. For that reason, the inhibition calculated was more than 100% (fig. 3).

During the infusion of low doses of AII, saralasin decreased RPF and suppressed RR ($-22 \pm 4\%$ and $-49 \pm 11\%$, respectively; n = 3). However, during the infusion of high doses of AII, saralasin increased RPF and did not significantly change RR ($+19 \pm 8\%$ and $-23 \pm 27\%$, respectively; n = 3). Two representative experiments are shown in figures 4 and 5.

In these and in the preceding experiments, it was observed that after the end of the simultaneous infusion of AII and saralasin in vasoconstrictor doses, RPF returned more slowly to control values than after the infusion of AII alone. 10 min after the end of an infusion of AII, RPF was not different from values measured before the start of the infusion ($+ 0.4$

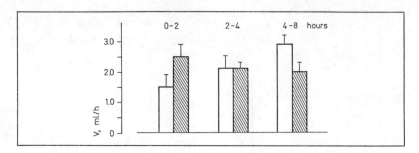

Fig. 6. Urine volume (V) 0–2, 2–4, and 4–8 h after glycerol injection (10 ml/kg) in rats during the infusion of saline (0.8 ml/8 h; white bars) or saralasin (10 µg/kg min, 0.8 ml/8 h; hatched bars) (means ± SE, n = 10 and 11, respectively).

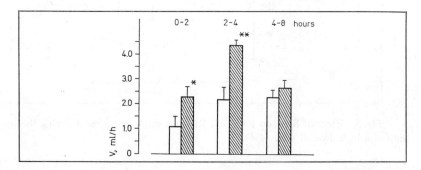

Fig. 7. Urine volume (V) 0–2, 2–4, and 4–8 h after glycerol injection (10 ml/kg) in rats during the infusion of rat serum alone (3.6 ml/4 h; white bars) or together with saralasin (10 µg/kg min, 0.8 ml/8 h; hatched bars) (means ± SE, n = 13 in both groups), * p<0.05; **p<0.001.

± 1.0%; n = 9), but 10 min after the end of an infusion of AII together with vasoconstrictor doses of saralasin RPF was still decreased (– 11 ± 3%; n = 10).

When saralasin was infused 5 min after the start of the perfusion in a dose (6–17 ng/ml), which inhibited the effect of infused AII by 66%, it did not significantly change RPF (– 3 ± 3%; n = 3).

Acute Renal Failure

No differences in urine volume were found between rats infused with saralasin and rats which received saline only (fig. 6). 24 h after the injection of glycerol, the plasma urea concentration was slightly lower in saralasin-treated rats than in controls, but the difference was of border-

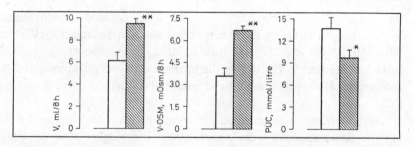

Fig. 8. Urine volume (V), solute excretion (VxOSM) and plasma urea concentration (PUC) 8 h after glycerol injection (10 ml/kg) in rats receiving rat serum only (3.6 ml/4 h) (white bars) or rat serum (3.6 ml/4 h) and saralasin (10 μg/kg min, 0.8 ml/8 h) together (hatched bars) (means ± SE, n = 28 for V and VxOSM, n = 13 for PUC in both groups). * $p<0.05$; ** $p<0.001$.

line statistical significance (22.5 ± 3.5 versus 26.8 ± 2.3 mmol/l, 0.05 $<p<0.1$, n = 10 and n = 11, respectively).

In the experiments with volume supplementation, urine volume was increased in saralasin-treated rats during the 4-hour period following glycerol injection (fig. 7). During the 8-hour period subsequent to the administration of glycerol, urine volume and solute excretion were significantly higher in rats which received saralasin than in controls, which had serum only (fig. 8). At 8 and 24 h after glycerol injection, plasma urea concentrations were less elevated in saralasin-treated rats than in the respective controls (8 h: 9.7 ± 1.0 versus 13.7 ± 1.5 mmol/l, $p<0.05$, n = 13 in both groups; 24 h: 19.5 ± 1.6 versus 25.7 ± 2.2 mmol/l, $p<0.05$, n = 15 in both groups).

During the period between 8 and 24 h after glycerol injection, urine volume did not significantly differ in saralasin-treated rats and controls which received serum only. When urine volume was calculated for the total 24-hour period after glycerol administration, the amounts were similar in saralasin-treated and control rats (17.1 ± 1.8 ml/24 h versus 16.9 ± 3.0 ml/24 h, n = 15 in both groups).

Discussion

Intrinsic Activity of AII Antagonists

Competitive AII antagonists may act at high concentrations as partial agonists [45, 73]. In the isolated perfused rat kidney, saralasin, in concentrations which effectively antagonize the renal vasoconstriction induced by AII, had no effect by its own on renal vascular resistance (RVR) and RR. VANDONGEN et al. [90], in their isolated kidney preparation, even observed an increase in RR during the infusion of low concentrations of AII antagonists. However, a similar rise occurred during the infusion of saline [91].

High concentrations of saralasin increased RVR and suppressed RR, but the doses required were about 1,000 times higher than equipotent doses of AII [33, 94]. These findings indicate that in the rat kidney the intrinsic activity of saralasin is low. Recently, it has also been reported that saralasin has no agonistic effect on renal blood flow in the intact rat [37].

During the infusion of high doses of saralasin, renal vasoconstriction has been observed in anaesthetized dogs [74] and also in rabbits on a high salt diet [55, 56]. In these studies, such an effect of saralasin was obtained only when plasma renin concentration (PRC) was low, and not when PRC was high [55, 74]. These observations concur with our results in the isolated kidney. High doses of saralasin increased RVR in the presence of low concentrations of AII, whereas similar or even higher doses of saralasin decreased RVR when AII concentrations in the perfusion medium were high.

The fact that the time required to revert to control values of RVR was longer after a vasoconstrictor infusion of saralasin than after an infusion of AII might be explained on the basis that Sar[1]-substituted analogues have a higher receptor-binding affinity and lower degradation rate [14, 29, 73].

Renal Haemodynamics and Circulating AII

In untreated dogs or rats, AII antagonists, infused intravenously or into the renal artery, did not change RVR [16, 18, 22, 37, 42, 49, 74]. Glomerular filtration rate and urinary sodium excretion were also not affected [16, 18, 49]. These data suggest that under resting conditions, circulating AII has no significant effect on renal haemodynamics.

When plasma renin concentration was elevated by acute renal artery constriction in the dog, saralasin increased blood flow in the contralateral, untouched kidney during intraarterial infusion [74, 93]. The degree of the saralasin-induced renal vasodilation was closely correlated with the arterial plasma renin levels [74].

Infusion of saralasin also increased renal blood flow in sodium-depleted rabbits and dogs, but changes in sodium excretion were either absent or small [22, 55]. Similar effects have been observed after inhibition of the converting enzyme by SQ 20881 [55, 92].

In dogs with chronic thoracic caval constriction, saralasin infused into the renal artery increased renal blood flow [22], but did not change urinary sodium excretion. In acute thoracic caval constriction, saralasin had no effect on renal blood flow although plasma renin levels were increased [80].

Renal blood flow also rose during the intrarenal infusion of saralasin in dogs with high output heart failure due to an arteriovenous fistula [23]. Since glomerular filtration rate did not increase, filtration fraction fell; urinary sodium excretion remained unchanged.

In the isolated perfused rat kidney [32], AII does not significantly decrease RPF at concentrations corresponding to plasma levels measured in the intact rat [64]. High concentrations of AII, comparable to plasma AII levels found after stimulation of the RAS, induce renal vasoconstriction [32]. This vasoconstrictor effect of infused AII can be completely blocked by saralasin [94].

Intrarenal Formation of AII

Numerous experimental data suggest that AII formed locally within the kidney may act on the renal arterioles and thereby determine RVR and glomerular filtration rate [86, 87]. The enzymes necessary for the formation and degradation of AII are present in the juxtaglomerular apparatus [27]; *in vivo*, renin substrate (RS) is continuously delivered to the kidney by the blood. RS might also be present within the kidney [41, 58]; moreover, AII has been demonstrated in renal tissue after flushing the kidney with saline [53]. Nonetheless, it remains to be shown that the intrarenal concentrations of RS suffice to produce functionally significant amounts of AII.

From former experiments, in which we infused RS, it may be inferred that AII is continuously formed from plasma RS within the kidney *in vivo* and that this reaction can be affected by the systemic administration of inhibitors of the RAS [34]. If the isolated kidney is perfused with an electrolyte solution free of RS, no AII should be formed. Accordingly, we have shown that neither saralasin nor SQ 20881, in amounts sufficient to blunt the effect of infused AI and AII, induced vasodilation [34]. Even when saralasin was given as early as 5 min after the start of the perfusion, RVR was unchanged. Our observations, that in the isolated kidney neither saralasin nor SQ 20881 did change RVR, suggest that in the absence of plasma RS the intrarenal formation of AII is either insignificantly low, or occurs at extravascular sites, which are not accessible to AII antagonists and converting enzyme inhibitors [34].

Autoregulation and Intrarenal AII

The increased delivery of sodium chloride to the distal tubule or the enhanced uptake of these ions by the macula densa cells is held to result in a reduction of single nephron filtration rate via a constriction of the afferent arteriole [59, 76, 78, 84]. Since increased distal tubular sodium load also activates renin in the juxtaglomerular apparatus [85], it has been inferred that the arteriolar vasoconstrictor response in the tubulo-glomerular feedback mechanism is mediated by the intrarenal RAS [86, 87].

SCHNERMANN and STOWE [77] observed that either $Me_2Gly^1\text{-}Ile^8$-AII or SQ 20881 reduced the vasoconstrictor feedback responses. However, tubulo-glomerular feedback was not completely blocked by these manoeuvres and other substances such as propranolol and xanthine derivatives had similar effects [77, 78].

The tubulo-glomerular feedback mechanism mediated by the intrarenal RAS has also been suggested to play an important role in the autoregulation of renal blood flow and filtration rate [83, 86]. However, neither $Sar^1\text{-}Gly^8$-AII nor $Sar^1\text{-}Ala^8$-AII reduced the renal capacity to autoregulate in the intact dog and in the isolated perfused dog kidney [2, 42, 43]. Similarly, SQ 20881 had no effect on autoregulation of renal blood flow and glomerular filtration in the dog [26, 30].

It has also been shown that autoregulation persists after renin depletion [25, 69] in the absence of plasma RS [95] and during the infusion of AII [7, 25, 46].

The negative results obtained with AII antagonists and converting enzyme inhibitors do not provide conclusive evidence against the participation of the RAS in autoregulation. It cannot be excluded that AII might be formed locally in concentrations which surmount those of the antagonists, or at sites which are inaccessible to the infused compounds [34, 87].

Renin Release

The negative feedback of circulating AII on RR is well established [9, 15, 89]. Since in the isolated perfused rat kidney, AII suppresses RR without inducing significant renal vasoconstriction, a direct effect of AII on the secretory mechanism of the epitheloid cells may be assumed [32].

AII antagonists can interfere with the negative feedback of AII on RR. Infusion of saralasin increased plasma renin levels in man [13, 67], dog [3, 23, 38, 39, 80], rabbit [81, 82] and rat [8, 44, 62, 82]. Comparable results have been obtained with SQ 20881 in man [28], dog [3, 54, 92] and rat [8].

Whereas an elevation of PRC by AII antagonists was not consistently observed in rats and in dogs with low PRC [3, 6, 8, 38, 39], saralasin or SQ 20881 generally induced a rise of PRC when the RAS was stimulated by renal artery constriction [3, 54], in malignant renal hypertension [40], after haemorrhage [24], salt deprivation [38, 39, 44, 81], adrenalectomy [8], thoracic caval constriction [38, 39, 80] or in high output cardiac failure [23].

It might be argued that the increase in RR during the administration of AII antagonists or SQ 20881 was a consequence of a fall in blood pressure or of an accompanying change in renal haemodynamics [17, 18], but in several of the studies mentioned RVR and blood pressure were unchanged [28, 62, 82, 92], or the fall in blood pressure was not correlated with the rise in PRC [13, 67, 82]. Hence, these studies with AII antagonists and converting enzyme inhibitors provide additional evidence for a direct suppression of RR by circulating AII.

Renal Prostaglandins

There is increasing evidence for a relationship between the RAS and renal prostaglandins [52]. On the one hand, AII infused into the kidney stimulates prostaglandin synthesis and release [50]. It was shown that prostaglandin release stimulated by AI or AII in the isolated rabbit kidney can be blocked by AII antagonists [60]. On the other hand, prostaglandins, acting as modulators, attenuate the renal vasoconstriction induced by AII [1, 51]; after blockade of prostaglandin synthesis the vasoconstrictor effect of AII may be unmasked.

When renal prostaglandin biosynthesis was blocked by either indomethacin or meclofenamate, renal vasoconstriction occurred [30, 47]. Such a response was not seen when saralasin was infused concomitantly [57, 75]. Hence the renal vasoconstriction after indomethacin administration seems to be, at least in part, dependent on basal levels of AII.

Acute Renal Failure

The observation of increased plasma renin levels [11, 12, 19, 20, 72, 88] and an enhanced activity of intrarenal renin [21] in several types of ARF supports the assumption that the RAS participates in the pathogenesis of this syndrome. However, AII antagonists failed to change plasma urea concentration or urine volume in rats 24 and 48 h after the induction of ARF by glycerol injection [4, 71] or temporary ischaemia [36]. Saralasin did not increase renal blood flow in dogs with unilateral ARF induced by the intraarterial infusion of norepinephrine [61].

In our experiments, we found that saralasin significantly improved renal excretory capacity in glycerol-induced ARF only when it was infused together with sufficient amounts of rat serum. From these observations, it may be concluded that a protective effect of saralasin in glycerol-induced ARF largely depends on concomitant volume substitution [5]. Therefore, the conflicting results obtained with passive immunization against AII in this type of ARF might, at least in part, be explained by the different amounts of serum administered [48, 63, 70, 72].

Our data do not provide information about the mechanism by which serum infusion enhanced the protective effect of saralasin. It is possible that in ARF intrarenal AII, which might be formed in high concentrations locally, is not effectively antagonized by saralasin. Serum administration might reduce the intrarenal activity of the RAS to such an extent [87] that a competitive antagonist, such as saralasin, becomes effective. Alternatively, serum administration might inhibit other vasoconstrictor systems activated in glycerol-induced ARF. Recently, we have found that in this type of ARF, plasma levels of vasopressin are 100-fold higher than in untreated rats [35]; in these concentrations, vasopressin might induce systemic and renal vasoconstriction. Thus, serum infusion might attenuate renal vasoconstriction by suppressing vasopressin release and thereby facilitating the action of an AII antagonist.

Our observation that saralasin only moderately improved solute excretion [5], as well as the negative findings of others [4, 36, 71], cannot be taken as evidence against a major role of the RAS in the pathogenesis of ARF. Since AII antagonists may inhibit intrarenally formed AII less than circulating AII, the administration of AII antagonists does not give quantitative information about the contribution of the RAS to the development of anuria.

Summary

In the isolated perfused rat kidney, saralasin inhibits the vasoconstrictor effect of angiotensin II in a dose-dependent manner. At high infusion rates saralasin, by itself, increases renal vascular resistance and suppresses renin release. Such an agonistic effect is not observed in the presence of high concentrations of angiotensin II.

In acute renal failure induced by glycerol, saralasin has a beneficial effect on urine volume, solute excretion and plasma urea concentration only when it is administered together with an adequate volume of rat serum.

Acknowledgements

We thank the Norwich Pharmacal Company (Norwich, N.Y.) for the gift of saralasin. We are indebted to Prof. Dr. E. HACKENTHAL for the determination of renin concentrations. These studies were supported by the German Research Foundation within the SFB 90 'Cardiovasculäres System'.

References

1 AIKEN, J. W. and VANE, J. R.: Intrarenal prostaglandin release attenuates the renal vasoconstrictor activity of angiotensin. J. Pharmac. exp. Ther. *184:* 678–687 (1973).
2 ANDERSON, R. J.; TAHER, M. S.; CRONIN, R. E.; McDONALD, K. M., and SCHRIER, R. W.: Effect of β-adrenergic blockade and inhibitors of angiotensin II and prostaglandins on renal autoregulation. Am. J. Physiol. *229:* 731–736 (1975).
3 AYERS, C. R.; VAUGHAN, E. D., jr.; YANCEY, M. R.; BING, K. T.; JOHNSON, C. C., and MORTON, C.: Effect of 1-sarcosine-8-alanine angiotensin II and converting enzyme inhibitor on renin release in dog acute renovascular hypertension. Circulation Res. *34/35:* I-27–33 (1974).
4 BARANOWSKI, R. L.; O'CONNOR, G. J., and KURTZMAN, N. A.: The effect of 1-sarcosine,8-leucyl angiotensin II on glycerol-induced acute renal failure. Archs int. Pharmacodyn. Thér. *217:* 322–331 (1975).
5 BAUEREISS, K.; HOFBAUER, K. G., and GROSS, F.: Effect of saralasin in glycerol-induced renal failure in rats; in GESSLER Proc. Symp. Berchtesgarden (Dustri, München, in press).
6 BECKERHOFF, R.; UHLSCHMID, G.; VETTER, W.; ARMBRUSTER, H.; NUSSBERGER, J.; RECK, G.; SCHMIED, U., and SIEGENTHALER, W.: Effect of angiotensin II and of an angiotensin II analogue (sar^1-ile^8-angiotensin II) on blood pressure, plasma aldosterone and plasma renin activity in the dog. Clin. Sci. mol. Med. *48:* 41s–44s (1975).
7 BELLEAU, L. J. and EARLEY, L. E.: Autoregulation of renal blood flow in the presence of angiotensin infusion. Am. J. Physiol. *213:* 1590–1595 (1967).

8   BING, J.: Rapid marked increase in plasma renin in rats treated with inhibitors of the renin system. Acta path. microbiol. scand. A *81:* 376–378 (1973).
9   BLAIR-WEST, J. R.; COGHLAN, J. P.; DENTON, D. A.; FUNDER, J. W.; SCOGGINS, J. A., and WRIGHT, R. D.: Inhibition of renin secretion by systemic and intrarenal angiotensin infusion. Am. J. Physiol. *220:* 1309–1315 (1971).
10 BOUCHER, R.; MÉNARD, J., and GENEST, J.: A micromethod for measurement of renin in the plasma and kidney of rats. Can. J. Physiol. Pharmacol. *45:* 881–890 (1967).
11 BROWN, J. J.; GLEADLE, R. I.; LAWSON, D. H.; LEVER, A. F.; LINTON, A. L.; MACADAM, R. F.; PRENTICE, E.; ROBERTSON, J. I. S., and TREE, M.: Renin and acute renal failure: studies in man. Br. med. J. *i:* 253–258 (1970).
12 BROWN, W. C. B.; BROWN, J. J.; GAVRAS, H.; JACKSON, A.; LEVER, A. F.; MACGREGOR, G.; MACADAM, R. F., and ROBERTSON, J. I. S.: Renin and acute circulatory renal failure in the rabbit. Circulation Res. *30:* 114–122 (1972).
13 BRUNNER, H. R.; GAVRAS, H.; LARAGH, J. H., and KEENAN, R.: Hypertension in man. Exposure of the renin and sodium components using angiotensin II blockade. Circulation Res. *34/35:* I-35–43 (1974).
14 BUMPUS, F. M. and KHOSLA, M. C.: Inhibition of the pressor and aldosterone-releasing effects of angiotensin II. Clin. Sci. mol. Med. *48:* 15s–18s (1975).
15 BUNAG, R. D.; PAGE, I. H., and MCCUBBIN, J. W.: Inhibition of renin release by vasopressin and angiotensin. Cardiovasc. Res. *1:* 67–73 (1967).
16 CADNAPAPHORNCHAI, P.; BOYKIN, J.; HARBOTTLE, J. A.; MCDONALD, K. M., and SCHRIER, R. W.: Effect of angiotensin II on renal water excretion. Am. J. Physiol. *228:* 155–159 (1975).
17 DAVIS, J. O.; FREEMAN, R. H.; JOHNSON, J. A., and SPIELMAN, W. S.: Agents which block the action of the renin-angiotensin system. Circulation Res. *34:* 279–285 (1974).
18 DAVIS, J. O.: The use of blocking agents to define the functions of the renin-angiotensin system. Clin. Sci. mol. Med. *48:* 3s–14s (1975).
19 DIBONA, G. F. and SAWIN, L. L.: The renin-angiotensin system in acute renal failure in the rat. Lab. Invest. *25:* 528–532 (1971).
20 FLAMENBAUM, W.; MCNEIL, J. S.; KOTCHEN, T. A., and SALADINO, A. J.: Experimental acute renal failure induced by uranyl nitrate in the dog. Circulation Res. *31:* 682–698 (1972).
21 FLAMENBAUM, W. and HAMBURGER, R. J.: The role of the renin-angiotensin system in the initiating phase of uranyl nitrate-induced acute renal failure in the rat. Kidney int. *6:* 42A (1974).
22 FREEMAN, R. H.; DAVIS, J. O.; VITALE, S. J., and JOHNSON, J. A.: Intrarenal role of angiotensin II. Homeostatic regulation of renal blood flow in the dog. Circulation Res. *32:* 692–698 (1973).
23 FREEMAN, R. H.; DAVIS, J. O.; SPIELMAN, W. S., and LOHMEIER, T. E.: High-output heart failure in the dog: systemic and intrarenal role of angiotensin II. Am. J. Physiol. *229:* 474–478 (1975).
24 FREEMAN, R. H.; DAVIS, J. O.; JOHNSON, J. A.; SPIELMAN, W. S., and ZATZMAN, M. L.: Arterial pressure regulation during hemorrhage: homeostatic role of angiotensin II. Proc. Soc. exp. Biol. Med. *149:* 19–22 (1975).

25 GAGNON, J. A.; KELLER, H. I.; KOKOTIS, W., and SCHRIER, R. W.: Analysis of role of renin-angiotensin system in autoregulation of glomerular filtration. Am. J. Physiol. *219:* 491–496 (1970).

26 GAGNON, J. A.; RICE, M. K., and FLAMENBAUM, W.: Effect of angiotensin converting enzyme inhibition on renal autoregulation. Proc. Soc. exp. Biol. Med. *146:* 414–418 (1974).

27 GRANGER, P.; DAHLHEIM, H., and THURAU, K.: Enzyme activities of the single juxtaglomerular apparatus in the rat kidney. Kidney int. *1:* 78–88 (1972).

28 HABER, E.; SANCHO, J.; RE, R.; BURTON, J., and BARGER, A. C.: The role of the renin-angiotensin-aldosterone system in cardiovascular homeostasis in normal man. Clin. Sci. mol. Med. 48: 49s–52s (1975).

29 HALL, M. M.; KHOSLA, M. C.; KHAIRALLAH, P. A., and BUMPUS, F. M.: Angiotensin analogs: the influence of sarcosine substituted in position 1. J. Pharmacol. exp. Ther. *188:* 222–228 (1974).

30 HERBACZYNSKA-CEDRO, K. and VANE, J. R.: Contribution of intrarenal generation of prostaglandin to autoregulation of renal blood flow in the dog. Circulation Res. *33:* 428–436 (1973).

31 HOFBAUER, K. G.; ZSCHIEDRICH, H.; RAUH, W., and GROSS, F.: Conversion of angiotensin I into angiotensin II in the isolated perfused rat kidney. Clin. Sci. *44:* 447–456 (1973).

32 HOFBAUER, K. G.; ZSCHIEDRICH, H.; HACKENTHAL, E., and GROSS, F.: Function of the renin-angiotensin system in the isolated perfused rat kidney. Circulation Res. *34/35:* I-193–201 (1974).

33 HOFBAUER, K. G.; ZSCHIEDRICH, H.; BAUEREISS, K., and GROSS, F.: Effects of angiotensin II and its antagonist saralasin on renal plasma flow and renin release in the isolated rat kidney. Pflügers Arch. *355:* R48 (1975).

34 HOFBAUER, K. G.; ZSCHIEDRICH, H., and GROSS, F.: Regulation of renin release and intrarenal formation of angiotensin. Studies in the isolated perfused rat kidney. Clin. exp. Pharmacol. Physiol. *3:* 73–93 (1976).

35 HOFBAUER, K. G.; BAUEREISS, K.; KONRADS, A.; HACKENTHAL, E.; MÖHRING, B.; MÖHRING, J., and GROSS, F.: Release of vasopressin and renin in glycerol-induced acute renal failure of rats. Pflügers Arch. *362:* R12 (1976).

36 IAINA, A.; SOLOMON, S., and ELIAHOU, H. E.: Reduction in severity of acute renal failure in rats by beta-adrenergic blockade. Lancet *ii:* 157–159 (1975).

37 ISHIKAWA, K. and HOLLENBERG, N. K.: Blockade of the systemic and renal vascular actions of angiotensin II with the 1-sar, 8-ala analogue in the rat. Life Sci. *17:* 121–130 (1975).

38 JOHNSON, J. A. and DAVIS, J. O.: Effects of a specific competitive antagonist of angiotensin II on arterial pressure and adrenal steroid secretion in dogs. Circulation Res. *32/33:* I-159–168 (1973).

39 JOHNSON, J. A. and DAVIS, J. O.: Angiotensin II: important role in the maintenance of arterial blood pressure. Science *179:* 906–907 (1973).

40 JOHNSON, J. A.; DAVIS, J. O.; SPIELMAN, W. S., and FREEMAN, R. H.: The role of the renin-angiotensin system in experimental renal hypertension in dogs. Proc. Soc. exp. Biol. Med. *147:* 387–391 (1974).

41 JOHNSTON, C. I.; MENDELSOHN, F. A. O.; HUTCHINSON, J. H., and MORRIS, B.:

Composition of juxtaglomerular granules isolated from rat kidney cortex; in SAMBHI Mechanisms of hypertension, pp. 238–248 (Excerpta Medica, Amsterdam 1973).

42 JOSE, P. A.; SLOTKOFF, L. M.; MONTGOMERY, S.; CALCAGNO, P. L., and EISNER, G.: Autoregulation of renal blood flow in the puppy. Am. J. Physiol. *229:* 983–988 (1975).

43 KALOYANIDES, G. J. and DIBONA, G. F.: The effect of 1-sarcosine-8-alanine-angiotensin II on function of the isolated dog kidney. Kidney int. *6:* 57A (1974).

44 KEETON, K. and PETTINGER, W.: Renin release induced by angiotensin antagonism: the effects of sodium balance and beta blockade. Fed. Proc. Fed. Am. Socs exp. Biol. *34:* 769 (1975).

45 KHOSLA, M. C.; SMEBY, R. R., and BUMPUS, F. M.: Structure-activity relationship in angiotensin II analogs; in PAGE and BUMPUS Handbook of experimental pharmacology, vol. 37, pp. 126–161 (Springer, Berlin 1974).

46 KIIL, F.; KJEKSHUS, J., and LØYNING, E.: Renal autoregulation during infusion of noradrenaline, angiotensin and acetylcholine. Acta physiol. scand. *76:* 10–23 (1969).

47 LONIGRO, A. J.; ITSKOVITZ, H. D.; CROWSHAW, K., and MCGIFF, J. C.: Dependency of renal blood flow on prostaglandin synthesis in the dog. Circulation Res. *32:* 712–717 (1973).

48 MATTHEWS, P. G.; MORGAN, T. O., and JOHNSTON, C. I.: The renin-angiotensin system in acute renal failure in rats. Clin. Sci. mol. Med. *47:* 79–88 (1974).

49 MCDONALD, K. M.; TAHER, S.; AISENBREY, G.; TORRENTE, A. DE, and SCHRIER, R. W.: Effect of angiotensin II and an angiotensin II inhibitor on renin secretion in the dog. Am. J. Physiol. *228:* 1562–1567 (1975).

50 MCGIFF, J. C.; CROWSHAW, K.; TERRAGNO, N. A., and LONIGRO, A. J.: Release of a prostaglandin-like substance into renal venous blood in response to angiotensin II. Circulation Res. *26/27:* I-121–130 (1970).

51 MCGIFF, J. C.; CROWSHAW, K.; TERRAGNO, N. A., and LONIGRO, A. J.: Renal prostaglandins: possible regulators of the renal actions of pressor hormones. Nature, Lond. *227:* 1255–1257 (1970).

52 MCGIFF, J. C. and ITSKOVITZ, H. D.: Prostaglandins and the kidney. Circulation Res. *33:* 479–488 (1973).

53 MENDELSOHN, F. A. O.: Direct evidence for local formation of angiotensin II in kidney. 6th Int. Congr. Nephrol., abstr. 262 (1975).

54 MILLER, E. D., jr.; SAMUELS, A. I.; HABER, E., and BARGER, A. C.: Inhibition of angiotensin conversion and prevention of renal hypertension. Am. J. Physiol. *228:* 448–453 (1975).

55 MIMRAN, A.; GUIOD, L., and HOLLENBERG, N. K.: The role of angiotensin in the cardiovascular and renal response to salt restriction. Kidney int. *5:* 348–355 (1974).

56 MIMRAN, A.; HINRICHS, K. J., and HOLLENBERG, N. K.: Characterization of smooth muscle receptors for angiotensin: studies with an antagonist. Am. J. Physiol. *226:* 185–190 (1974).

57 MIMRAN, A.; CASELLAS, D., and BARJON, P.: Effect of 1-sarcosine-8-

alanine-angiotensin II on the renal hemodynamic changes induced by indomethacin in the rat. 6th Int. Congr. Nephrol., abstr. 180 (1975).

58 MORRIS, S. J. and JOHNSTON, C. I.: Renin substrate in granules from rat kidney cortex. Biochem. J. *154:* 625–637 (1976).

59 MÜLLER-SUUR, R.; GUTSCHE, H.-U.; SAMWER, K. F.; OELKERS, W., and HIERHOLZER, K.: Tubuloglomerular feedback in rat kidneys of different renin contents. Pflügers Arch. *359:* 33–56 (1975).

60 NEEDLEMAN, P.; KAUFFMAN, A. H.; DOUGLAS, J. R., jr.; JOHNSON, E. M., jr., and MARSHALL, G. R.: Specific stimulation and inhibition of renal prostaglandin release by angiotensin analogs. Am. J. Physiol. *224:* 1415–1419 (1973).

61 NEWHOUSE, J. H. and HOLLENBERG, N. K.: Vascular characteristics of unilateral acute renal failure in the dog. Investve Radiol. *9:* 241–251 (1974).

62 OATES, H. F.; STOKES, G. S., and GLOVER, R. G.: Plasma renin response to acute blockade of angiotensin II in the anaesthetized rat. Clin. exp. Pharmacol. Physiol. *1:* 155–160 (1974).

63 OKEN, D. E.; COTES, S. C.; FLAMENBAUM, W.; POWELL-JACKSON, J. D., and LEVER, A. F.: Active and passive immunization to angiotensin in experimental acute renal failure. Kidney int. *7:* 12–18 (1975).

64 OSTER, P.; HACKENTHAL, E., and HEPP, R.: Radioimmunoassay of angiotensin II in rat plasma. Experientia *29:* 353–354 (1973).

65 OSTER, P.; BAUKNECHT, H., and HACKENTHAL, E.: Active and passive immunization against angiotensin II in the rat and rabbit. Evidence for a normal regulation of the renin-angiotensin system. Circulation Res. *37:* 607–614 (1975).

66 PALS, D. T.; MASUCCI, F. D.; DENNING, G. S., jr.; SIPOS, F., and FESSLER, D. C.: Role of the pressor action of angiotensin II in experimental hypertension. Circulation Res. *29:* 673–681 (1971).

67 PETTINGER, W. A. and MITCHELL, H. C.: Renin release, saralasin and the vasodilator-beta-blocker drug interaction in man. New Engl. J. Med. *292:* 1214–1217 (1975).

68 PETTINGER, W. A.; KEETON, K., and TANAKA, K.: Radioimmunoassay and pharmakokinetics of saralasin in the rat and hypertensive patients. Clin. Pharmac. Ther. *17:* 146–158 (1975).

69 POTKAY, S. and GILMORE, J. P.: Autoregulation of glomerular filtration in renin-depleted dogs. Proc. Soc. exp. Biol. Med. *143:* 509–513 (1973).

70 POWELL-JACKSON, J. D.; LEVER, A. F.; MACADAM, R. F.; ROBERTSON, J. I. S.; BROWN, J. J.; MACGREGOR, J.; TITTERINGTON, D. M., and WAITE, M. A.: Protection against acute renal failure in rats by passive immunization against angiotensin II. Lancet *i:* 774–776 (1972).

71 POWELL-JACKSON, J. D.; MACGREGOR, J.; BROWN, J. J.; LEVER, A. F., and ROBERTSON, J. I. S.: The effect of angiotensin II antisera and synthetic inhibitors of the renin-angiotensin system on glycerol-induced acute renal failure in the rat; in FRIEDMAN and ELIAHOU Proc. Conf. Acute Renal Failure, DHEW publ., NIH 74-608, pp. 281–289 (1973).

72 RAUH, W.; OSTER, P.; DIETZ, R., and GROSS, F.: The renin-angiotensin system in acute renal failure of rats. Clin. Sci. mol. Med. *48:* 467–473 (1975).

73 Regoli, D.; Park, W. K., and Rioux, F.: Pharmacology of angiotensin. Pharmac. Rev. *26:* 69–123 (1974).
74 Satoh, S. and Zimmerman, B. G.: Effect of (Sar^1, Ala^8)angiotensin II on renal vascular resistance. Am. J. Physiol. *229:* 640–645 (1975).
75 Satoh, S. and Zimmerman, B. G.: Influence of the renin-angiotensin system on the effect of prostaglandin synthesis inhibitors in the renal vasculature. Circulation Res. *36/37:* I-89–96 (1975).
76 Schnermann, J.; Wright, F. S.; Davis, J. M.; Stackelberg, W. v., and Grill, G.: Regulation of superficial nephron filtration rate by tubulo-glomerular feedback. Pflügers Arch. *318:* 147–175 (1970).
77 Schnermann, J. and Stowe, N.: The role of the renin-angiotensin and adrenergic systems in the mediation of tubulo-glomerular feedback. Pflügers Arch. *347:* R 66 (1974).
78 Schnermann, J.: Regulation of filtrate formation by feedback. Proc. 6th Int. Congr. Nephrol. (Karger, Basel, in press, 1976).
79 Schurek, H. J.; Brecht, J. P.; Lohfert, H., and Hierholzer, K.: The basic requirement for the function of the isolated cell-free perfused rat kidney. Pflügers Arch. *354:* 349–365 (1975).
80 Slick, G. L.; DiBona, G. F., and Kaloyanides, G. J.: Renal blockade to angiotensin II in acute and chronic sodium-retaining states. J. Pharmacol. exp. Ther. *195:* 185–193 (1975).
81 Steele, J. M., jr. and Lowenstein, J.: Differential effects of an angiotensin II analogue on pressor and adrenal receptors in the rabbit. Circulation Res. *35:* 592–600 (1974).
82 Stokes, G. S.; Oates, H. F., and Weber, M. A.: Angiotensin blockade in studies of the feedback control of renin release in rats and rabbits. Clin. Sci. mol. Med. *48:* 33s–36s (1975).
83 Thurau, K.: Renal hemodynamics. Am. J. Med. *36:* 698–719 (1964).
84 Thurau, K. und Schnermann, J.: Die Natriumkonzentration in den Macula densa-Zellen als regulierender Faktor für das Glomerulumfiltrat (Mikropunktionsversuche). Klin. Wschr. *43:* 410–413 (1965).
85 Thurau, K. W. C.; Dahlheim, H.; Grüner, A.; Mason, J., and Granger, P.: Activation of renin in the single juxtaglomerular apparatus by sodium chloride in the tubular fluid at the macula densa. Circulation Res. *30:* II-182–186 (1972).
86 Thurau, K.: Intrarenal action of angiotensin; in Page and Bumpus Handbook of experimental pharmacology, vol. 37, pp. 475–489 (Springer, Berlin 1974).
87 Thurau, K. and Mason, J.: The intrarenal function of the juxtaglomerular apparatus; in Guyton and Thurau Kidney and urinary tract physiology, pp. 357–389 (Butterworth, London/ University Park Press, Baltimore 1974).
88 Tu, W. H.: Plasma renin activity in acute tubular necrosis and other renal diseases associated with hypertension. Circulation *31:* 686–695 (1965).
89 Vander, A. J. and Geelhoed, G. W.: Inhibition of renin secretion by angiotensin II. Proc. Soc. exp. Biol. Med. *120:* 399–403 (1965).
90 Vandongen, R.; Peart, W. S., and Boyd, G. W.: Effect of angiotensin II and

its nonpressor derivatives on renin secretion. Am. J. Physiol. *226:* 277–283 (1974).
91 VANDONGEN, R. and PEART, W. S.: The inhibition of renin secretion by alpha-adrenergic stimulation in the isolated rat kidney. Clin. Sci. mol. Med. *47:* 471–479 (1974).
92 VAUGHAN, E. D., jr.; KIMBROUGH, H. M., jr.; CAREY, R. M., and AYERS, C.: Intrarenal role of angiotensin II for control of glomerular filtration and renal blood flow during sodium restriction in conscious dogs. 6th Int. Congr. Nephrol., abstr. 487 (1975).
93 ZIMMERMAN, B. G.: Involvement of angiotensin-mediated renal vasoconstriction in renal hypertension. Life Sci. *13:* 507–515 (1973).
94 ZSCHIEDRICH, H.; HOFBAUER, K. G.; HACKENTHAL, E.; BARON, G. F., and GROSS, F.: Intrarenal formation of angiotensin II in the rat: interference by saralasin and SQ 20881. Clin. Sci. mol. Med. *48:* 37s–40s (1975).
95 ZSCHIEDRICH, H.; HOFBAUER, K. G.; BARON, G. D.; HACKENTHAL, E., and GROSS, F.: Relationship between perfusion pressure and renin release in the isolated rat kidney. Pflügers Arch. *360:* 255–266 (1975).

Dr. K. G. HOFBAUER, Department of Pharmacology, University of Heidelberg, Im Neuenheimer Feld 366, *D–6900 Heidelberg* (FRG)

Discussion[1]

BOYD: Dr. BLAIR-WEST, I was very impressed with your data on P113. However, in these days when receptors are viewed as rather dynamic entities, I wonder if they might change in sodium depletion. I think that most of your studies where you gave angiotensin, and overcame that effect with saralasin, were done in the sodium-replete state. Have you looked at mild to moderate sodium depletion on the basis that the receptor might there change to have a higher affinity for angiotensin than saralasin and so not be so well blocked by this antagonist?

BLAIR-WEST: No, we have not done that. One of the reasons is that we do not find a potentiated response to exogenous angiotensin in sodium-deficient animals. In fact, the moderately sodium-deficient sheep is virtually insensitive to exogenous angiotensin II. Therefore, in the system we are testing here, that would not really be a test because there is no control response.

POULSEN: In connection with these excellent papers, I would like to suggest that we consider the effect of angiotensin and antagonists in the same way as we usually do it with other hormones and receptors. That is: (1) the angiotensin is bound to the receptor; (2) when bound, it has an agonistic effect. Since all peptide analogues have at least some agonistic properties, it would be helpful in interpreting the results if dose-response curve were plotted as agonistic effect versus log dose for both angiotensins and antagonists. This would be very helpful in evaluating what a high and a low dose meant. In fact, all so-called 'inhibitors' are just agonists with a dose-response curve shifted more or less to the right.

BLAIR-WEST: On the point raised by Dr. POULSEN, we have tried to explore that agonist relationship for P113 and other antagonists too, but we do not have much evidence that there is a simple dose-response relation from our data so far. We certainly do not have as much P113 data as we have data for angiotensin II; with that, we have good evidence for dose-response relationships. We have not found any suggestion of a parallel dose-response curve for P113. It is interesting

1 Discussion of the papers of BLAIR-WEST et al. and HOFBAUER et al.

that the biggest doses of antagonists we have used in sodium-replete sheep have never produced more than a 2-µg/h aldosterone response. I guess that means we just have not gone far enough, but then I do not think Norwich would give us enough of the analogue!

VANDONGEN: Dr. HOFBAUER, I was rather surprised that you were unable to block the inhibitory effect of angiotensin on renin release considering the very high doses of P113 you used. I am wondering whether you have some way of assessing the potency or in fact, the purity of the batches of P113?

HOFBAUER: I do not think that this was a problem. What I wanted to show you in these experiments was that agonistic and antagonistic effects of saralasin depend on the dose of the antagonist as well as on the concentration of angiotensin II. We have also performed experiments in which saralasin antagonized the renin suppression induced by angiotensin II.

FUNDER: Dr. BLAIR-WEST, were there any changes in adrenal blood flow over that very wide gamut of concentrations that you infused, because that might be an index of a vascular competition for angiotensin-binding sites?

BLAIR-WEST: The answer to that is no – there was never any effect on adrenal blood flow.

FUNDER: Secondly, just to bring things back to the aldosterone control area, were there any changes in glucocorticoid secretion rates?

BLAIR-WEST: No. At the doses we infused, there was no consistent effect at all on glucocorticoid secretion; but there is one interesting point, the rate of infusion that we used in sodium-replete animals tended to lower plasma potassium concentration in adrenal venous plasma. This may have had some impact on the effect of P113 infused in sodium-replete sheep.

FUNDER: Dr. POULSEN, referring to the response curves you mentioned, have you any evidence that the plateau of the agonist activity of a partial agonist is at the same height as that of AII, for instance?

POULSEN: No. I did not elaborate this, but I am suggesting that if dose-responce curves were presented for angiotensin and antagonist it might be easier to interpret the results.

PETTINGER: I am quite fascinated, Dr. BLAIR-WEST, by your report showing the quantitative difference in the blockade by saralasin between the endogenous and the exogenous sources of angiotensin. We had the same experience using a different model. Angiotensin when infused (exogenous), stimulated aldosterone which was in turn completely blocked by saralasin. Conversely, endogenous angiotensin mediation of aldosterone release (consisting of minoxidil-induced renin release) could not be blocked as effectively by saralasin [CAMPBELL et al., J. Pharmac. exp. Ther., 1975].

In Stokes and Edwards: Drugs Affecting the Renin-Angiotensin-Aldosterone System. Use of Angiotensin Inhibitors
Prog. biochem. Pharmacol., vol. 12, pp. 86–97 (Karger, Basel 1976)

Effects of Angiotensin Antagonists in Various Forms of Experimental Arterial Hypertension

C. M. Ferrario[1], F. M. Bumpus, Z. Masaki, M. C. Khosla and J. W. McCubbin

Research Division, Cleveland Clinic Foundation, Cleveland, Ohio

Contents

Introduction	86
Methods	87
Results	88
Effect of Angiotensin Antagonists on Renovascular Hypertension	88
Findings in Chronic Hypertension Due to Cellophane Perinephritis	90
Effect of Angiotensin Blockers in Malignant Hypertension	91
Discussion	92
Summary	95
Acknowledgments	95
References	96

Introduction

The synthesis of specific angiotensin antagonists [14, 15, 18] has provided a tool to assess the role that the renin-angiotensin system plays in renal hypertension. Initial studies, as reviewed recently by Davis *et al.* [5], have not been conclusive, in part because compounds with various degrees of competitive properties were used and, more often than not, anesthesia was employed during measurements. To overcome these problems, we have resorted to techniques previously developed in this laboratory [10] for serial measurements of arterial pressure and cardiac output

---

1   Dr. C. M. Ferrario is an Established Investigator of the American Heart Association.

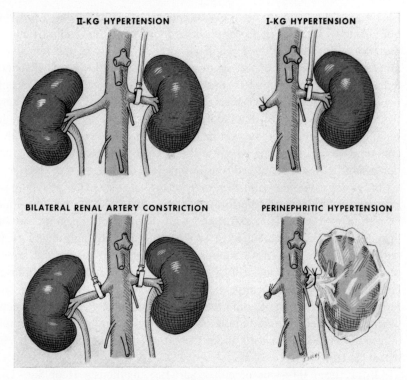

Fig. 1. The four forms of experimental renal hypertension used to study the effect of the angiotensin inhibitor on the elevated arterial pressure.

in conscious dogs. The two most potent competitive antagonists of the naturally occurring angiotensin II peptide (Sar^1-Ile^8- and Sar^1-Thr^8-angiotensin II) were given repeatedly to conscious renal hypertensive dogs to elucidate the importance of the renin-angiotensin system in the evolution of various forms of renal hypertension. It was considered necessary to study multiple types of renal hypertension because others have suggested [2] that the renin-angiotensin system participates in some, but not all forms of renal hypertension.

Methods

Four models of renal hypertension were produced in a group of 32 conscious dogs (fig. 1). In 23, experimental *renovascular hypertension* was produced by constriction of one renal artery with (13 dogs) and without (10 dogs) removal of the opposite kidney. While one-kidney renal hypertension can be produced readily in dogs by reducing the lumen of a renal artery to about 50% of its original diameter [9], a somewhat more tedious process is required to produce in dogs a model of renovascular hypertension with an intact normal kidney. This process involves a newly described technique [16] that incorporates a two-step occlusion of the left renal artery: an initial reduction in renal blood flow stimulates growth of collateral vessels to a degree sufficient to maintain adequate renal function when the main renal artery is completely occluded 14 days later. In six other dogs, *perinephritic hypertension* was produced by wrapping one kidney in cellophane and performing a contralateral nephrectomy. Finally, a preparation regularly associated with the prompt appearance of malignant hypertension and a most striking increase in the activity of the renin-angiotensin system was obtained by *constricting both renal arteries severely* [12]. In all dogs, arterial pressure was recorded from an indwelling catheter in an iliac artery; measurements were made daily in a quiet room and with the dogs resting quietly on a soft pad. Plasma renin activity was measured by radioimmunoassay every second day from samples of venous blood. Details of these techniques have been described elsewhere [7, 10].

Intravenous infusions of either Sar^1-Ile^8-angiotensin II or Sar^1-Thr^8-angiotensin II (0.075–4.0 μg/kg/min) were given with a small electrolytically driven pump connected to a catheter chronically inserted into a jugular vein and worn on the dog's leather vest from 1 to 48 h. Synthetic angiotensin II (0.5–1.5 μg) and norepinephrine (10 μg) dissolved in saline and injected intravenously as boluses were given repeatedly during the studies. On other occasions, saline instead of the angiotensin antagonists was infused for the same periods of time.

Results

Effect of Angiotensin Antagonists on Renovascular Hypertension

Constriction of one renal artery in conjunction with contralateral nephrectomy (one-kidney Goldblatt model) or without contralateral nephrectomy (two-kidney Goldblatt model) is accompanied by a rapid rise in arterial pressure. In dogs, this is somewhat more pronounced in the one-kidney model [7, 10]. In both models of renovascular hypertension,

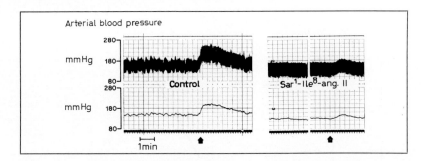

Fig. 2. Effect of Sar[1]-Ile[8]-angiotensin II on the elevated blood pressure of a one-kidney renal hypertensive dog three days after constriction of a sole remaining renal artery. Left panel: Phasic (top) and mean arterial pressure (bottom) in a trained conscious dog prior to infusion of the angiotensin inhibitor. Middle panel: Blood pressure response following a 2-hour intravenous infusion of the blocker at a dose of 0.60 µg/kg/min. Angiotensin II (1.0 µg i.v. bolus) given at arrows.

plasma renin activity increases significantly above the values established before renal artery constriction to reach an elevated plateau at about the 4th to 6th day after operation. After this time, plasma renin activity in both forms of hypertension gradually returns to control values but at a slightly more pronounced rate in dogs with one-kidney hypertension. Renin activity is consistently normal within 12–17 days after renal artery constriction whether or not the contralateral intact kidney is present [10].

Infusions of a competitive angiotensin antagonist within *one to five days* after constriction of one renal artery induced a significant blood pressure drop in both the one- and two-kidney animals. The fall in blood pressure lasted for the duration of the infusion and was associated with a marked reduction in the pressor response to injected angiotensin II (fig. 2).

In dogs with one-kidney renal hypertension, 5-hour infusions of Sar[1]-Ile[8]-angiotensin II at a rate of 0.200 µg/kg/min caused mean arterial pressure to fall from a group average of 147 ± 4 to 117 ± 2 mm Hg and plasma renin activity to decline by an average of $27 \pm 9\%$ ($p<0.01$) below the elevated values found prior to infusion (5.0 ± 3.0 ng/ml/h). In dogs with two-kidney hypertension, treated one to five days after renal artery occlusion, decreases in arterial pressure averaged 10 ± 4 mm Hg ($p<0.05$) but unlike animals with one-kidney hypertension, the hypotensive effect of the angiotensin analogue was associated with an increase rather than a decrease in plasma renin activity. In this group of animals, plasma renin activity rose from 6 ± 2 to 10 ± 4 ng/ml/h ($p<0.01$).

The hypotensive effects of the angiotensin antagonists seen in the

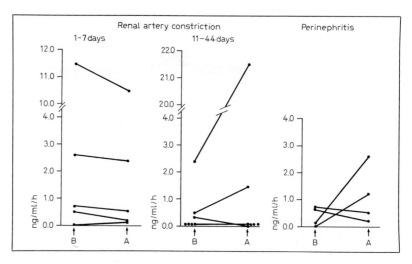

Fig. 3. Changes in plasma renin activity before (arrow B) and after (arrow A) infusion of Sar^1-Ile^8-angiotensin II in a group of dogs with one-kidney renal hypertension.

early phase of one- and two-kidney renal hypertension could not be elicited again as the hypertension evolved into a more chronic phase. Infusions of either Sar^1-Ile-^{8}- or Sar^1-Thr^8-angiotensin II at a rate of 1.0 µg/kg/min or even higher rates (4.0 µg/kg/min) had no significant hypotensive effects. Moreover, in most of these animals (18 of 23 dogs), the angiotensin analogue caused rises in blood pressure that ranged from 10 to 24 mm Hg and were associated with various alterations in plasma renin activity. In dogs with one renal artery constricted and the opposite kidney removed, the concentration of renin in plasma did not change significantly in five and increased in two others (fig. 3). Similar results were obtained in dogs with chronic two-kidney hypertension.

Thus, these experiments appear to indicate that angiotensin II is prominently involved in the early phase of both one- and two-kidney models of renal hypertension but that its role becomes less obvious as hypertension progresses into a more chronic phase.

Findings in Chronic Hypertension Due to Cellophane Perinephritis

At the time of experiment, hypertension due to cellophane perinephritis had been sustained for from 65 days to seven months and mean arterial pressure had stabilized at values that were always above 150 mm Hg (range: 155–200 mm Hg). Plasma renin levels prior to the infusion of

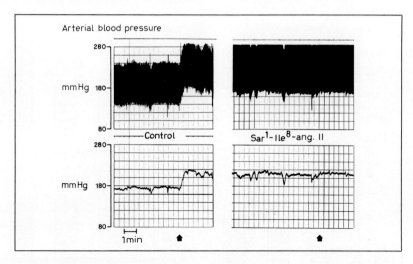

Fig. 4. Agonistic effects of Sar[1]-Ile[8]-angiotensin II in a renal hypertensive dog three months after development of cellophane perinephritis. Angiotensin II (1.0 μg) given as a bolus intravenously at arrows. The hypertensive response (right panel) was recorded 45 min after commencing the infusion.

the antagonists were entirely normal, usually less than 1.0 ng/ml/h. Continuous infusion of either Sar[1]-Ile[8] or Sar[1]-Thr[8]-angiotensin II for from 5 to 48 h elicited marked pressor responses that were sustained for about 45 min following the start of the infusion (fig. 4). After this time, arterial pressure returned to the previously elevated control levels where it remained for the duration of the infusion. This pressor response probably reflects a direct action of the angiotensin antagonists on vascular smooth muscle since both analogues had about the same pressor effects and only Sar[1]-Ile[8]-angiotensin II stimulates release of catecholamines from the adrenal medulla [19]. Changes in plasma renin activity as a result of the infusion were variable and not related to the effect of the angiotensin inhibitor on blood pressure (fig. 3).

Effect of Angtiotensin Blockers in Malignant Hypertension

The pathogenesis of necrotizing vascular lesions that develop in a fulminating manner during the course of experimental and clinical hypertension has intrigued us for a number of years [11, 12]. In dogs, malignant hypertension is characterized by severe hypertension, high plasma concentration of renin and rapid loss of renal function. The administration of angiotensin antagonists at doses between 1 and 2 μg/kg/min had a

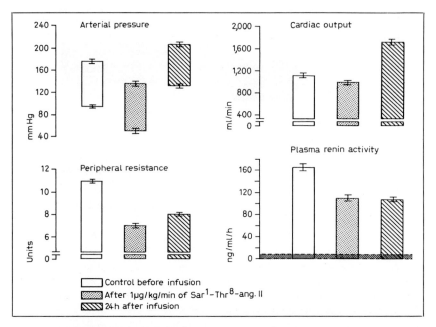

Fig. 5. Effect of Sar[1]-Thr[8]-angiotensin II on the evolution of malignant hypertension due to severe constriction of the renal arteries (5 dogs). Bars are means +SE of systolic and diastolic arterial pressure, cardiac output (electromagnetic flowmeter), peripheral resistance and plasma renin activity before, after 125 min of infusion (1 μg/kg/min) and on the day after. Shaded baseline represents normal range of plasma renin activity in normal dogs.

profound effect on the elevated blood pressure of conscious dogs with accelerated hypertension. Arterial blood pressure could be reduced markedly (fig. 5) because of a direct fall in peripheral resistance without significant change in cardiac output. The fall in blood pressure was accompanied by decreases in plasma renin activity that persisted 24 h after stopping the infusion (fig. 5).

Discussion

There is still uncertainty concerning the extent of involvement of the renin-angiotensin system in experimental and clinical forms of renal hypertension. There appears to be agreement only in that increased circulating levels of renin are present for a few days after production of experimental hypertension; the activity of the enzyme in blood appears to be normal with hypertension of longer duration. Normal levels of circulating renin in sustained forms of both human and experimental renal hypertension have led many to believe that factors other than the renin-angiotensin system are responsible for maintaining the elevated blood pressure. Certainly, this latter hypothesis has been somewhat substantiated by the demonstration that, more often than not, both passive and active immunization against angiotensin II have failed either to prevent or decrease the elevated blood pressure of renal hypertensive animals; for review, see CARRETERO and BUJAK [4] and DAVIS et al. [5]. These observations are also in accord with clinical studies where it has been shown that peripheral plasma renin activity is normal in a 'disturbingly high percentage of hypertensive patients who have subsequently benefited from operations for renovascular hypertension' [20]. These data have made it difficult to understand how hypertension can be attributed only to excessive levels of angiotensin II.

The synthesis of peptides with competitive properties against angiotensin II has provided us with more precise tools for evaluating the role of the renin-angiotensin system in renal hypertension since these peptides are capable of inhibiting completely all of the cardiovascular actions of angiotensin II [21, 22].

Both Sar^1-Ile^8- and Sar^1-Thr^8-angiotensin II have been shown to be most potent competitive antagonists of the parent hormone; in addition, the latter inhibitor of angiotensin II has been shown by PEACH and ACKERLY [19] to have no agonistic activity in the adrenal medulla. Therefore, the present experiments should have allowed for an adequate evaluation of the role of the renin angiotensin system in arterial hypertension of renal origin. The observations of SWEET et al. [21] on the effect of Sar^1-Ile^8-angiotensin II upon the immediate hypertensive response that follows renal artery constriction prompted us to extend these preliminary observations to conscious hypertensive dogs.

We found that continuous infusion of these angiotensin blockers produced a partial lowering of blood pressure only during the acute phase of one- and two-kidney renal hypertension. At this early stage of hypertension, part of the rise in arterial pressure therefore may be renin-dependent. On the other hand, the response of the elevated blood pressure to additional infusions of the blockers became negligible as hypertension progressed to a more chronic phase, at a time when the concentration of renin in peripheral plasma was not above normal values. Likewise, these angiotensin antagonists had no effect on the elevated blood pressure of dogs with hypertension due to perinephritis, induced by cellophane wrapping, an established method leading to a more chronic and benign form of renal hypertension.

It has been advocated recently [1] that two different mechanisms operate to maintain the elevated blood pressure of one- and two-kidney renal hypertension. Because administration of an angiotensin inhibitor (Sar^1-Ala^8-angiotensin II) to rats with either one- or two-kidney renal hypertension of six weeks duration caused significant blood pressure drops only in those animals with an intact contralateral kidney, they postulated that the two-kidney model was renin-dependent and the other was not. Such an important conclusion was made, however, without taking the precaution of measuring plasma renin levels before and after administration of the angiotensin II inhibitor. GAVRAS *et al.* [13] re-studied the problem by infusing the angiotensin inhibitor in two-kidney renal hypertensive rats with hypertension lasting for 15 weeks instead of 6. The inhibitor failed to reduce arterial pressure at this time. A decrease in blood pressure could be obtained when the animals were sodium-depleted by a combination of salt restriction and intravenous injection of furosemide.

These results, consistent with those reported previously by us [3, 22] indicate that two-kidney hypertension in its chronic phase is similar to that in which the contralateral kidney has been removed. BRUNNER and GAVRAS [2] have interpreted these results as follows: during the chronic phases of both one- and two-kidney hypertension, arterial pressure is maintained by a mechanism other than the renin-angiotensin system. Because salt restriction causes no changes in resting blood pressure but restores the ability of the angiotensin II inhibitor to lower the elevated blood pressure of two-kidney hypertensive rats, arterial pressure at this stage may be dependent on fluid volumes. The basic difference between the two models of experimental renal hypertension may reside then solely in an effect of the elevated arterial pressure on the function of the contralateral normal kidney. As the intact kidney loses its ability to handle sodium chloride, apparent differences in pathogenesis between the two models of experimental renal hypertension are no longer present. As appealing as this hypothesis may be, it still requires confirmation. When this hypothesis is examined more closely the sodium-volume explanation does not suffice. Plasma and extracellular fluid volumes are always normal, or slightly below normal [7, 10], in dogs with various forms of chronic renal hypertension, and this is also a feature of most patients with established essential hypertension [6]. We believe that these data argue against the possibility that chronic arterial hypertension associated with normal renin values is attributable to volume expansion. It could be that the transition from a renin-dependent phase to one that is not may be accounted for on the basis of a primary neurogenic abnormality that becomes fully developed as hypertension progresses into a more chronic phase. The many aspects of this hypothesis were detailed by Page [17] as early as 1949.

Severe hypertension accompanied by high levels of circulating renin and associated vascular lesions resembling nephrosclerosis in man can now be produced readily in dogs [12]. The competitive inhibitors of angiotensin II had a profound effect on the elevated arterial pressure of dogs with malignant hypertension, lowered peripheral resistance and decreased plasma renin activity. The fall in the latter was not as pronounced as the effect on blood pressure and appeared to persist even after the arterial pressure had returned again to values well within those encountered prior to infusion. This lack of parallelism between the fall in both arterial pressure and plasma renin activity is yet to be understood, but in view of what we now know it appears to suggest a less than 'solo' role for the renin angiotensin system in the etiology of malignant hypertension. This view has been presented previously [12].

Summary

The dependence of renal hypertension on increased levels of angiotensin II was investigated in conscious dogs at various stages of hypertension of four different types. Two of the most potent angiotensin inhibitors, Sar^1-Ile^8- and Sar^1-Thr^8-angiotensin II, were infused separately in various doses and at different times throughout the evolution of renal hypertension. The results of these experiments are consistent with the view that the renin-angiotensin system may participate in the acute and malignant phases of renal hypertension; they do not provide evidence for its participation when hypertension enters the chronic phase.

Acknowledgments

We wish to express our gratitude to Dr. CHARLES SWEET for his contribution to parts of this research. The authors wish to acknowledge the contribution of Dr. EMMANUEL BRAVO, Director of the Radioimmunoassay Laboratory.

References

1 BRUNNER, H. R.; KIRSHMAN, J. D.; SEALEY, J. E., and LARAGH, J. H.: Hypertension of renal origin. Evidence for two different mechanisms. Science *174:* 1344–1346 (1971).
2 BRUNNER, H. R. and GAVRAS, H.: The role of renin and sodium in high blood pressure regulation; in BERGLUND, HANSSON and WERKO Pathophysiology and management of arterial hypertension, pp. 32–42 (Lindgren & Soner, Molndal 1975).
3 BUMPUS, F. M.; SEN, S.; SMEBY, R. R.; SWEET, C.; FERRARIO, C. M., and KHOSLA, M. C.: Use of angiotensin II antagonists in experimental hypertension. Circulation Res. *32–33:* suppl. I, pp. 150–158 (1973).
4 CARRETERO, O. A. and BUJAK, B.: The role of angiotensin in benign and malignant experimental hypertension: and immunological approach; in GENEST and KOIW Hypertension '72, pp. 473–481 (Springer, Berlin 1972).
5 DAVIS, J. O.; FREEMAN, R. H.; JOHNSON, J. A., and SPIELMAN, W. S.: Agents which block the action of the renin-angiotensin system. Circulation Res. *34:* 279–285 (1974).
6 DOYLE, A. E.: Sympathetic nervous activity in hypertension. Hosp. Pract. *10:* 87–95 (1975).
7 FERRARIO, C. M.; PAGE, I. H., and MCCUBBIN, J. W.: Increased cardiac output as a contributory factor in experimental renal hypertension. Circulation Res. *27:* 799–810 (1970).
8 FERRARIO, C. M.; BLUMLE, C.; NADZAM, G. R., and MCCUBBIN, J. W.: An externally adjustable renal artery clamp. J. appl. Physiol. *31:* 635–637 (1971).
9 FERRARIO, C. M. and MCCUBBIN, J. W.: Renal blood flow and pressure before and after development of renal hypertension. Am. J. Physiol. *224:* 102–110 (1973).
10 FERRARIO, C. M.: Contribution of cardiac output and peripheral resistance to experimental renal hypertension. Am. J. Physiol. *226:* 711–717 (1974).
11 FERRARIO, C. M. and MCCUBBIN, J. W.: The unresolved nature of malignant hypertension. Acta physiol. latinoam. *24:* 569–573 (1974).
12 FERRARIO, C. M.; HELMCHEN, U. M., and MCCUBBIN, J. W.: On the differences and similarities between the benign and malignant phases of renal hypertension; in MILLIEZ and SAFAR Recent advances in hypertension, vol. 2, pp. 159–169 (Boehringer Ingelheim, Reims 1975).
13 GAVRAS, H.; BRUNNER, H. R.; THURSTON, H., and LARAGH, J. H.: Reciprocation of renin dependency with sodium volume dependency in renal hypertension. Science *188:* 1316–1317 (1975).
14 KHAIRALLAH, P. A.; TOTH, A., and BUMPUS, F. M.: Analogs of angiotensin II. Mechanism of receptor interaction. J. Med. Chem. *13:* 181–184 (1970).
15 MARSHALL, G. R.; VINE, W., and NEEDLEMAN, P.: Specific competitive inhibitor of angiotensin II. Proc. natn. Acad. Sci. USA *67:* 1624–1630 (1970).
16 MASAKI, Z.; FERRARIO, C. M.; BUMPUS, F. M., and KHOSLA, M. C.: Is two-kidney hypertension renin dependent? Fed. Proc. Fed. Am. Socs exp. Biol. *35:* 556 (1976).

17 Page, I. H.: Pathogenesis of arterial hypertension. J. Am. med. Ass. *10:* 451–457 (1949).
18 Pals, D. T.; Masucci, F. D.; Sipos, F., and Denning, G. S., jr.: Specific competitive antagonist of the vascular action of angiotensin II. Circulation Res. *29:* 664–672 (1971).
19 Peach, M. J. and Ackerly, J.: Angiotensin antagonists in adrenal cortex and medulla. Fed. Proc. Fed. Am. Socs exp. Biol. (in press, 1977).
20 Streeten, D. H. P.; Anderson, G. H.; Freiberg, J. M., and Dalakos, T. G.: Use of an angiotensin II antagonist (saralasin) in the recognition of 'angiotensinogenic' hypertension. New Engl. J. Med. *292:* 657–662 (1975).
21 Sweet, C. S.; Ferrario, C. M.; Khosla, M. C., and Bumpus, F. M.: Antagonism of peripheral and central effects of angiotensin II by [Sar^1,Ile^8] angiotensin II. J. Pharmac. exp. Ther. *185:* 35–41 (1973).
22 Sweet, C. S.; Ferrario, C. M.; Kosoglov, A., and Bumpus, F. M.: Cardiovascular evaluation of [Sar^1,Ile^8] angiotensin II and its effects on blood pressure of conscious renal hypertensive dogs. in Wesson and Fanelli Recent advances in renal physiology and pharmacology, pp. 257–268 (University Park Press, Baltimore 1974).

C. M. Ferrario, MD, Research Division, Cleveland Clinic Foundation, 9500 Euclid Avenue, *Cleveland, OH 44106* (USA)

Antagonists, Inhibitors and Antisera in the Evaluation of Vascular Renin Activity

The Role of Local Generation of Angiotensin II

J. D. SWALES

University Department of Medicine, General Hospital, Leicester

Contents

Introduction	98
Vascular Wall Renin Activity	99
Changes in Vascular Renin	99
Role of Vascular Renin Activity	100
Vascular Renin and Angiotensin Antisera	101
Pressor Responsiveness to A II and Sodium Balance	104
AII Antagonist versus Antiserum	105
Timing of Pressor Changes after Nephrectomy	109
Response to AII Antagonist and Inhibitor after Nephrectomy	109
Conclusions	110
Summary	111
References	111

Introduction

In the long-standing discussion of the role of vascular renin in blood pressure control, there have been three main logical steps. The first of these involved the observation that renin-like activity was present in the blood vessel wall: this observation is now generally accepted. The second step of the analysis demands the demonstration that this renin-like activity changes in response to physiological stimuli. The third step requires proof that there are changes in the local generation of angiotensin II (AII) which exert a physiologically significant effect.

Vascular Wall Renin Activity

Earlier studies indicated that tissue from the arterial wall contained an enzyme which released a pressor substance from plasma, although the identity of the enzyme was uncertain [1, 2]. GOULD et al. [3] were able to show that this enzyme could not be distinguished from renin by a wide range of kinetic studies. In the tissue studied (abdominal aorta) renin-like activity was present to the greatest extent in the adventitia and the outer media. Vascular renin activity has subsequently been demonstrated in other species and other arteries [4, 5]. Less is known about the other components of the renin system necessary for the local generation of AII. However, converting enzyme has been demonstrated in the arterial bed of the hind limb of the dog by comparing the vasoconstrictor response to angiotensin I infusion before and after administering converting enzyme inhibitor [6]. Such activity is also present in the hind limb of the sheep [7] and in the mesenteric ciculation [8]. COLLIER and ROBINSON [9] have produced evidence that this is of functional importance in man. Angiotensin I (AI) was infused into the brachial artery and hand vein; it produced reduction in forearm blood flow and hand vein size, which could be prevented by infusion of converting enzyme inhibitor. The action of AI was too rapid to be caused by plasma conversion and these workers conclude that conversion was occurring in the vein wall.

Changes in Vascular Renin

Less is known about the changes which occur in the renin-angiotensin system of the blood vessel wall in different physiological states. However, ROSENTHAL et al. [5] found that rat aortic and plasma renin activity (PRA) fell in parallel following bilateral nephrectomy, although the half-life of aortic renin activity was substantially greater than that of PRA. Conversely, a rise in both plasma and aortic renin activity occurred in response to water deprivation and bilateral adrenalectomy. The same group later studied renin activity in the mesenteric artery of dogs. An increase in renin activity at this site occurred in response to congestive cardiac failure, sodium restriction and acute aldosterone administration, whilst chronic administration of aldosterone, progesterone and hydrochlorothiazide lowered it [4]. A slight non-significant increase in arterial renin activity occurred in animals with hypertension due to unilateral ischaemia whether the opposite kidney was in situ or not. However, hypertension in these animals was of four month's duration and may therefore not have been renin-mediated. Curiously, arterial renin activity was elevated in dogs subjected to the Goldblatt two-kidney procedure when these animals had failed to develop hypertension. Mesenteric renin activity did not fall in the 12 days after bilateral nephrectomy, and the authors conclude it is more likely to be formed locally. Whether the differences in behaviour of vascular renin activity after nephrectomy represent a species difference or whether they are related to the different vascular sites sampled is not clear. This is a matter of some importance since, if vascular renin activity is to have the functional importance which our group has attributed to it, then the renin has to have its origin in the kidney.

Role of Vascular Renin Activity

Most of the workers who have published observations upon vascular renin activity have speculated on its role in the control of peripheral resistance and hence of blood pressure. More recently, a possible analogy between noradrenaline and AII has been proposed. It is postulated that each agent acts as a local hormone in addition to having a more general action through secretion into the circulation (by the adrenal medulla in one case and the kidney in the other [10]). McGiff and Vane [11] have postulated a local vascular system whereby a pressor (including renin) and antipressor (including prostaglandins) system of hormones are each active in peripheral vascular control. Such hypotheses demand demonstration that vascular renin results in local generation of AII which plays a role in blood pressure control. Alternatively, it is possible that renin is either passively trapped or actively formed in the blood vessel wall but is inactive because of a deficiency in one component of the renin-angiotensin system in tissue adjacent to receptor sites. It is our contention that there is a group of experimental phenomena which can only be explained by the local activity of AII, generated in vasoactive quantities within the blood vessel wall [12–17].

Schaechtelin et al. [18] used an isovolaemic cross circulation preparation for the bioassay of renin in bilaterally nephrectomised rats. Circulation between the two rats was established after the injection of renin. Data presented by these workers suggested that the pressor effect of renin was maintained in donor rats beyond the point at which circulating material could be detected by the indicator animal. BING and NIELSON [19] later also demonstrated a pressor effect of injected renin which persisted well beyond the point at which such renin had disappeared from the circulation.

The first experiments in which specific physiological changes were attributed to vascular renin activity were those of DAUM et al. [20]. These workers found that amino-peptidase infusion diminished the pressor response of pithed rats to AII infusion: the impairment of the pressor response to renin injection was much less conspicuous. It was concluded that locally generated AII within the blood vessel wall was less susceptible to destruction by aminopeptidase.

Vascular Renin and Angiotensin Antisera

Our interest in vascular renin activity arose from quite a different source – this was the need to postulate generation of vasoactive quantities of AII at a site inaccessible to antibody in order to explain several curious *in vivo* properties of angiotensin antisera. BRUNNER et al. [21] measured the volume of AII antiserum needed to block the pressor response to a standard dose of AII in rats. They found that where PRA was elevated (i.e. after salt depletion) relatively small amounts of antibody were required to exert a blocking action. On the other hand, where PRA was low (i.e. after salt loading or bilateral nephrectomy), substantially larger volumes of antibody were required. This group concluded that the affinity of vascular receptors was conditioned by the state of salt balance; they further suggested that the relationship between receptor affinity and sodium balance was abnormal in hypertension. We were able to confirm (fig. 1) that dose response curves relating the volume of antibody to the pressor response to AII were moved to the right by salt loading and to the left by salt depletion [13]. However, the infusion of small amounts of renin also moved the curve sharply to the left. This relationship was also a quantitative one, so that it was possible to define a dose-response relationship between the quantity of renin infused and the requirement of antibody needed to block the animal against a standard dose of AII [13]. There was also a highly significant positive correlation between the response to a standard dose of AII and the volume of antibody needed to block the animal against the pressor effect of AII [12]. Saturation of antibody-binding sites by circulating AII clearly was not the factor which determined the

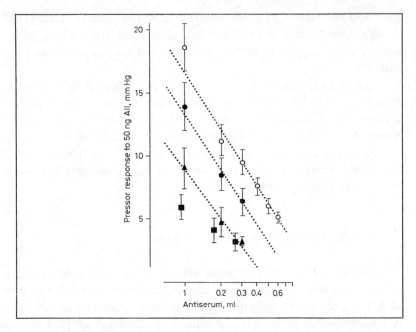

Fig. 1. Pressor response to a 50-ng bolus of AII after 0.1 ml doses of AII antiserum. Open circles = high salt pretreated rats; closed circles = normal rats; triangles = salt-depleted rats; squares = high salt pretreated renin-infused rats.

blocking dose of antiserum since less antiserum was required when circulating renin levels were high and vice versa. Further, the dramatic change in blocking requirement induced by renin infusion suggested that vascular receptor occupancy by AII was of critical importance. The only hypothesis which could reconcile these apparent contradictions requires the postulate that renin generated vasoactive quantities of AII at a site which was relatively inaccessible to the antibody molecule (fig. 2). In high renin states, therefore, vascular AII receptors would be largely occupied by AII which could not be bound by antibody; hence, vascular responsiveness to exogenous AII was low and the amount of antibody required to reduce free-circulating AII below the threshold for a pressor response was small. This would further explain the close correlation between blocking requirement for antiserum and the pressor response to AII, since both these variables would depend upon receptor occupancy by endogenous AII. Support for the view that conversion of AI to AII took place at a site which was inaccessible to antibody was obtained by OATES and

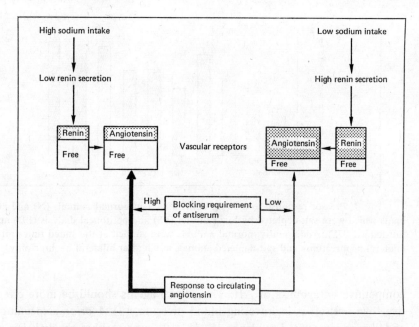

Fig. 2. Schematic view of occupation of vascular AII receptors in salt-loaded and salt-depleted animals.

STOKES [22]. This group found that AII antibody was less effective in blocking intraarterial than in blocking intravenous AI and that the pressor response to intravenous AI was much less well blocked than the pressor response to intravenous AII.

The hypothesis that generation of AII occurs at an antibody-inaccessible site in the blood vessel wall and that occupation of vascular AII receptors by locally generated AII is of major importance in the control of peripheral resistance makes possible several predictions that can be tested experimentally; these are: (1) Removal of the major source of renin by nephrectomy or blockade of AII formation should eliminate the difference in pressor response to AII produced by changes in sodium balance. Further, if vascular renin activity originates by local secretion rather than by uptake of circulating renin, only converting enzyme inhibition should eliminate differences in reactivity. (2) Since the vascular AII receptor site is accessible to circulating AII, it should be nearly equally accessible to molecules of similar size and configuration, in particular to

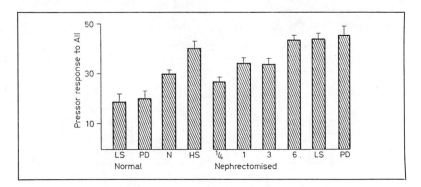

Fig. 3. Pressor response to a 50-ng bolus of AII in normal animals (N) and animals which were salt-depleted by low salt diet (LS) and peritoneal dialysis (PD) and salt-loaded (HS). Additionally, normal animals were studied at the stated times after bilateral nephrectomy and salt-depleted animals at 6 h after bilateral nephrectomy.

competitive antagonists of AII. Thus, these agents should be more effective than antiserum in lowering blood pressure. (3) *A priori* expectation and the above quoted evidence [5, 18, 19] suggests that the half-life of vascular renin is much greater than that of circulating renin. Changes in pressor responsiveness to AII after bilateral nephrectomy should therefore occur over a much longer period than would be predicted from the relatively short half-life of plasma renin. (4) On the same basis, it would be anticipated that where hypertension is partially or wholly due to renin hypersecretion, blood pressure should fall more slowly after bilateral nephrectomy than could be explained by the fall in plasma renin. Likewise, the blood pressure lowering action of angiotensin antagonists and inhibitors should be preserved even after plasma renin levels have fallen to the low levels encountered after bilateral nephrectomy. All the predictions have been examined experimentally over the last three years.

Pressor Responsiveness to AII and Sodium Balance

We have examined the effect of changes in sodium balance upon the pressor response to AII in the presence and absence of kidneys in order to determine whether changes in sodium balance affect vascular receptors directly or whether changes in pressor responsiveness to AII are renin-mediated. Prolonged and acute sodium depletion was induced in groups of rats (6 in each group) in two ways, by dietary sodium restriction for 8–14 days and by peritoneal dialysis using a solution of low sodium content [12]. This produced the anticipated fall in the pressor response to a standard dose of AII. Dietary salt loading of animals increased the pressor response. After nephrectomy, the pressor response was increased to the same degree whether the animal was sodium depleted or not (fig. 3). The volume of antiserum needed to block the pressor response to a standard dose of AII showed closely parallel [12] changes. The kidney therefore plays an essential role in determining the changes in pressor responsiveness to AII produced by changes in sodium balance. That this effect is mediated by endogenous AII formed as a result of renin secretion was shown by the use of a converting enzyme inhibitor [15]. When this was administered to salt depleted rats, the pressor response to AII became equal to that observed in nephrectomised and high salt pretreated animals. It is thus reasonable to conclude that the changes in pressor responsiveness to exogenous AII seen with changes in sodium balance are due to changes in the occupation of vascular receptors by endogenous AII. Whilst this observation taken alone does not exclude a role for circulating AII resulting in receptor occupancy, the changes in antibody-blocking requirement can only be explained by AII formed at an antibody inaccessible site.

AII Antagonist versus Antiserum

Several groups have studied the efficacy of active or passive immunization against AII in lowering the blood pressure of rats with hypertension produced by unilateral renal ischaemia with the opposite kidney in place (Goldblatt two-kidney hypertension).

Conclusions have been conflicting although even where a fall in blood pressure occurred animals did not become normotensive [23–27]. Immunization against AI has also proved unsuccessful in lowering the blood pressure [28]. When hypertension was induced in bilaterally nephrectomised rats by renin infusion, BING and NIELSON [19] observed that animals infused with AII antiserum showed less of a depressor response than animals infused with angiotensin antagonist. We have compared the success of antiserum and antagonist in reducing blood pressure in these two models of hypertension [16, 17]. In these studies, first antiserum then antagonist were administered through a jugular cannula and di-

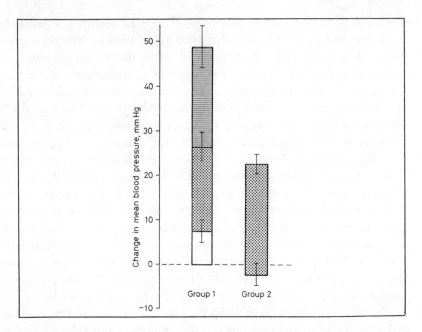

Fig. 4. Response to AII antibody and antagonist infusion in renin-induced hypertension. Height of columns indicates response to renin, cross-hatched part of the column represents the antibody-induced blood pressure fall, and the dotted part of the column the fall in blood pressure with antagonist. In group 1, renin was infused before antibody, and in group 2 afterwards. In each group, the antagonist produced a further fall in blood pressure.

rect blood pressure monitored from the carotid artery. Antiserum in quantities sufficient to block the pressor effect of a 50-ng bolus of AII was followed by infusion of sarcosine-alanine-AII (10 μg/min). Four groups of animals were investigated: (1) Nine bilaterally nephrectomised rats with acute hypertension produced by infusion of 0.2 U of MRC standard porcine renin. (2) A second group of 11 rats in which renin was infused in the interval between the antiserum and antagonist infusions. (3) Eight rats with unilateral renal artery constriction after 7–10 days of hypertension. (4) Seven rats, with hypertension of 4–6 months' duration, produced by a similar procedure.

In group 1, blocking doses of antiserum produced a partial fall in blood pressure of rats with renin-induced hypertension. A further substantial fall occurred when antagonist was infused (fig. 4) and blood pressure returned to near normal values. The renin-induced blood pressure

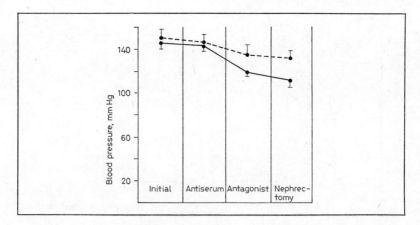

Fig. 5. Blood pressure changes in Goldblatt two-kidney hypertension induced by AII antibody, antagonist and nephrectomy. --- = Chronic; —— = short-term.

rise was limited in the antiserum pretreated animals of group 2, and antagonist then reduced blood pressure to normal levels (fig. 4). The animals remained blocked to exogenous AII throughout the procedure. A similar pattern of changes was observed in the rats with short term Goldblatt 2 hypertension (group 3) where antiserum produced only a minor non-significant fall in blood pressure whilst antagonist substantially lowered it (fig. 5). The effect of antagonist was less in animals with longstanding Goldblatt two-kidney hypertension (group 4). In all these experiments, therefore, in the same animals on the same occasion, antagonist was markedly more successful in lowering blood pressure than antiserum. OSTER *et al.* [29] have suggested that this comparative impotence of AII antisera is due to high levels of physiologically active free-circulating AII even in the presence of excess antibody. The main evidence quoted in favour of this hypothesis is that renin secretion is normal in passively immunized animals. It is argued that sufficient free AII must be present to preserve feedback inhibition of renin secretion. However, these observations are equally consistent with the view that renin secretion is inhibited by locally generated AII at a site in the juxtaglomerular apparatus which is inaccessible to antiserum. This explanation would seem more likely in view of the fact that the juxtaglomerular AII receptors are special examples of vascular AII receptors. There are three further arguments which militate against the hypothesis that high free levels of circulating AII maintain blood pressure even in the presence of excess AII antibody.

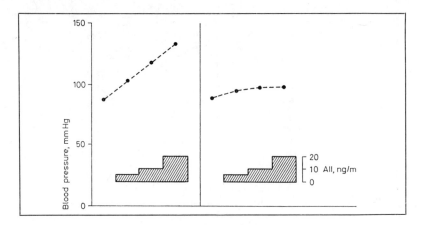

Fig. 6. Response of bilaterally nephrectomised rats to AII infusion, before, and after infusion of 0.4 ml of AII antiserum.

1) The blood pressure response to a bolus injection of 50 ng of AII was blocked in our experiments. Whilst OSTER *et al.* [29] maintain that only rapidly changing concentrations of AII may be 'buffered' by antibody, it is noteworthy that where protein binding of a compound is known to neutralise its vascular action, i.e. in diazoxide therapy, rapid bolus injections are vasoactive whilst no vascular action occurs when steady state conditions are established. This is therefore precisely the opposite pattern of response to that predicted by OSTER *et al.* [29].

2) Antiserum is effective against hypertension produced by infusion of AII (fig. 6.)

3) Irreversible antagonism of AII produces similar phenomena to AII antisera. Thus, amino-peptidase infusion is much more effective in preventing a pressor response to AII infusion than to renin infusion [20]. The hypothesis advanced by OSTER *et al.* [29] therefore seems to us less likely than the explanation postulated here.

Timing of Pressor Changes after Nephrectomy

In view of the evidence for endogenous renin activity as a determinant of vascular AII responsiveness, such changes in responsiveness after bilateral nephrectomy should parallel the degree of occupation of vascular AII receptors by locally generated AII. If circulating AII is the relevant factor, responsiveness should change in parallel with the fall in plasma renin, i.e. with a half-life of 10–16 min [30, 31]. In fact, the change in pressor responsiveness whether measured by pressor effect of graded doses of AII (fig. 3) or by the volume of antiserum needed to block a standard dose of AII, only reaches a maximum at 6 h after bilateral nephrectomy [12]. Further, STREWLER *et al.* [32] were able to demonstrate that prior salt restriction of the donor rabbit impaired the response of an *in vitro* preparation of aortic strip to AII. Clearly circulating renin cannot be the relevant factor in this situation.

Response to AII Antagonist and Inhibitor after Nephrectomy

The fall in blood pressure produced by sarcosine-alanine-AII in Goldblatt two-kidney hypertension occurs within a minute or two of the compound reaching the circulation once the angiotensin-dependent mechanism is inhibited; therefore, the response is prompt. It would be predicted that removal of renin by bilateral nephrectomy could cause a fall in blood pressure which paralleled the fall of circulating renin, if this variable was the relevant controlling factor. However, the fall in blood pressure occurs over a much longer period with only a minute fall in the first hour after bilateral nephrectomy (fig. 7).

In order to confirm that maintenance of blood pressure in this situation was due to persistence of renin outside the circulation, we infused

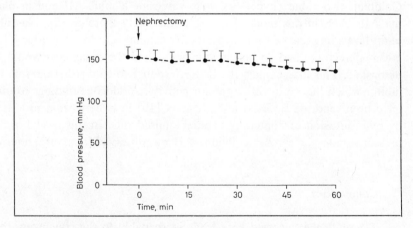

Fig. 7. Fall of blood pressure of rats with Goldblatt two-kidney hypertension for the first hour after bilateral nephrectomy.

Table I. Changes in blood pressure of rats with Goldblatt two-kidney hypertension (mean ± SEM). Converting enzyme inhibitor was infused at varying intervals after bilateral nephrectomy

Time after bilateral nephrectomy, h	Number of animals	Change in blood pressure mm Hg
1	9	−38.0 ± 10.1
2	8	−32.1 ± 5.7
6	9	−6.5 ± 2.7
1 + sarcosine-alanine AII	10	+3.8 ± 3.0

250 µg of converting enzyme inhibitor (SQ20881) into rats with two kidney Goldblatt hypertension at varying intervals after bilateral nephrectomy. A marked depressor action was observed at 1 and 2 h and a smaller blood pressure fall at 6 h (table I). That this fall was due to converting enzyme inhibition rather than the bradykinin-potentiating action of the compound was confirmed by prior infusion of 1 µg/min of sarcosine-alanine-AII. This successfully blocked the depressor response to converting enzyme inhibitor observed 1 h after bilateral nephrectomy.

Sarcosine-alanine-AII itself produced a significant fall in blood pressure (14.5 ± 3.6 mm Hg). This compound was less satisfactory in this experimental situation, however, as occasionally it produced a marked pressor effect when used in doses of 10 µg/min and even occasionally at the lower dose of 1 µg/min. The cause of this is probably the agonist action of the compound. There is considerable heterogeneity in the response of Goldblatt two-kidney hypertension to sarcosine-alanine-AII and in plasma AII levels in this model [33]. It seems likely therefore that vascular renin levels in some of these hypertensive animals are low 1 h after bilateral nephrectomy and so the pressor activity of the antagonist would be unmasked. This may explain the better response to converting enzyme inhibitor which has no intrinsic pressor activity. Analogous findings to ours have been reported by BING and NIELSON [19] in hypertension produced by renin infusion of bilaterally nephrectomised rats. In this model, a depressor response to sarcosine-alanine-AII was observed for several hours.

Conclusions

The evidence reviewed here leads inescapably to the conclusion that endogenous production of AII from renin maintains blood pressure and determines vascular responsiveness to exogenous AII even after circulating renin has fallen to insignificant levels. This hypothesis further allows for an explanation of some of the anomalous results obtained in studies using the infusion of AII antisera and also accounts for the different effects of antiserum and antagonist infusion. At the same time, a role is provided for the vascular renin activity which has been repeatedly demonstrated by earlier workers.

Summary

It is well established that the arterial wall contains renin-like material. It is suggested that this renin generates significant quantities of angiotensin II (AII) which occupy vascular AII receptors. This is important in determining the pressor response to exogenous AII and in the maintenance of blood pressure. Four lines of evidence support this view. (a) Prevention of AII generation by nephrectomy or converting enzyme inhibition increases the pressor response to AII which then becomes independent of sodium balance. (b) AII antagonists are more effective in renin-mediated hypertension than AII antisera. (c) Changes in renin-mediated pressor responsiveness to AII after bilateral nephrectomy take a much longer time to occur than can be explained in terms of changes in circulating renin activity: these changes are also observed in isolated vessels. (d) The depressor response to antagonist and inhibitors is preserved after bilateral nephrectomy for much longer periods than can be accounted for by circulating levels of renin.

References

1   DENGLER, H.: Über einen Reninartigen Wirkstoff in Arterienextrakten. Arch. exp. Path. Pharmakol. *227:* 481–487 (1956).
2   JIMÉNEZ-DIAZ, C.; BARREDA, P. DE LA y MOLINA, A. F.: La regulación química de la presión arterial. Revta clin. esp. *24:* 417–419 (1947).
3   GOULD, A. B.; SKEGGS, L. T., and KAHN, J. R.: The presence of renin activity in blood vessel wall. J. exp. Med. *119:* 389–399 (1964).
4   HAYDUK, K.; GANTEN, D.; BOUCHER, R., and GENEST, J.: in GENEST and KOIW Arterial and urinary renin activity. Hypertension 1972, pp. 435–443 (Springer, Berlin 1972).
5   ROSENTHAL, J.; BOUCHER, R.; ROJO-ORTEGA, J. M., and GENEST, J.: Renin activity in aortic tissue of rats. Can. J. Physiol. Pharmacol *47:* 53–56 (1969).

6   AIKEN, J. W. and VANE, J. R.: Inhibition of converting enzyme of the renin-angiotensin system in kidney and hindlegs of dogs. Circulation Res. *30:* 263–273 (1972).
7   OSBORN, E. C.; TILDESLEY, G.; O'GORMAN, L. P., and MAHLER, R. F.: The effect of angiotensin I and II on hind-limb blood flow in sheep. J. Pharm. Pharmac. *23:* 466–468 (1971).
8   DISALVO, J. and MONTEFUSCO, C. B.: Conversion of angiotensin I to II in the canine mesenteric circulation. Am. J. Physiol. *221:* 1576–1579 (1971).
9   COLLIER, J. G. and ROBINSON, B. F.: Comparison of effects of locally infused angiotensin I and II on hand veins and forearm arteries in man: evidence for converting enzyme activity in limb vessels. Clin. Sci. mol. Med. *47:* 189–192 (1974).
10 VANE, J. R.: Sites of conversion of angiotensin I; in GENEST and KOIW Hypertension '72, pp. 523–532 (Springer, Berlin 1972).
11 McGIFF, J. C. and VANE, J. R.: Prostaglandins and the regulation of blood pressure. Kidney int. *8:* S262–S270 (1975).
12 SWALES, J. D.; TANGE, J. D., and THURSTON, H.: Vascular angiotensin II receptors and sodium balance in rats: role of kidneys and vascular renin activity. Circulation Res. *37:* 96–100 (1975).
13 SWALES, J. D. and THURSTON, H.: Generation of angiotensin II at peripheral vascular level: studies using angiotensin II antisera. Clin. Sci. mol. Med. *45:* 691–700 (1973).
14 SWALES, J. D.; THURSTON, H., and TANGE, J. D.: Vascular angiotensin II in hypertension. Evidence for local generation of angiotensin. 6th Int. Congr Nephrology, Florence 1975, p. 553.
15 THURSTON, H. and LARAGH, J. H.: Prior receptor occupancy as a determinant of the pressor activity of infused angiotensin II in the rat. Circulation Res. *36:* 113–117 (1975).
16 THURSTON, H. and SWALES, J. D.: Action of angiotensin antagonist and antiserum upon the pressor response to renin: further evidence for the local generation of angiotensin II. Clin. Sci mol. Med. *46:* 273–276 (1974).
17 THURSTON, H. and SWALES, J. D.: Comparison of angiotensin II antagonist and antiserum infusion with nephrectomy in the rat with two-kidney Goldblatt hypertension. Circulation Res. *35:* 325–329 (1974).
18 SCHAECHTELIN, G.; REGOLI, D., and GROSS, F.: Quantitative assay and disappearance rate of circulating renin. Am. J. Physiol. *206:* 1361–1364 (1964).
19 BING, J. and NIELSON, K.: Cause of the prolonged pressor action of renin in nephrectomized rats. Acta path. Microbiol. A *81:* 247–253 (1973).
20 DAUM, A.; UEHLEKE, H. und KLAUS, D.: Unterschiedliche Beeinflussung der Blutdruckwirkung von Renin und Angiotensin durch Aminopeptidase. Arch. Pharmakol. exp. Path. *254:* 327–333 (1966).
21 BRUNNER, H. R.; CHANG, P.; WALLACH, R.; SEALEY, J. E., and LARAGH, J. H.: Angiotensin II vascular receptors: their avidity in relationship to sodium balance, the autonomic nervous system and hypertension. J. clin. Invest. *51:* 58–67 (1972).
22 OATES, H. F. and STOKES, G. S.: Role of extrapulmonary conversion in mediat-

ing the systemic pressor activity of angiotensin I. J. exp. Med. *140:* 79–85 (1974).
23 BING, J. and POULSEN, K.: Effect of anti-angiotensin on blood pressure and sensitivity to angiotensin and renin. Acta pathol. microbiol. scand A *78:* 6–18 (1970).
24 BRUNNER, H. R.; KIRSHMAN, J. D.; SEALEY, J. E., and LARAGH, J. H.: Hypertension of renal origin: evidence for two different mechanisms. Science *174:* 1344–1346 (1971).
25 CARRETERO, O. A.; KUK, P.; PIWONSKA, S.; HOULE, J. A., and MARIN-GREZ, M.: Role of the renin-angiotensin system in the pathogenesis of severe hypertension in rats. Circulation Res. *29:* 654–663 (1971).
26 HEDWALL, P. R.: Effect of rabbit antibodies against angiotensin II on the pressor response to angiotensin II and renal hypertension in the rat. Br. J. Pharmacol. *34:* 623–629 (1968).
27 MACDONALD, G. J.; BOYD, G. W., and PEART, W. S.: Renal hypertension and angiotensin antibodies. Am. Heart J. *83:* 137–139 (1972).
28 OATES, H. F.; STOKES, G. S.; STOREY, B. G.; GLOVER, R. G., and SNOW, B. F.: Renal hypertension in rats immunized against angiotensin I and angiotensin II. J. exp. Med. *139:* 239–248 (1974).
29 OSTER, P.; BAUKNECHT, H., and HACKENTHAL, E.: Active and passive immunization against angiotensin II in the rat and rabbit: evidence for a normal regulation of the renin-angiotensin system. Circulation Res. *37:* 607–614 (1975).
30 OATES, H. F.; FRETTON, J. A., and STOKES, G. S.: Disappearance rate of circulating renin after bilateral nephrectomy in the rat. Clin. exp. Pharmacol. Physiol. *1:* 547–549 (1974).
31 PETERS-HAEFELI, L.: Rate of inactivation of endogenous and exogenous renin in normal and in renin-depleted rats. Am. J. Physiol. *221:* 1339–1345 (1971).
32 STREWLER, G. J.; HINRICHS, K. J.; GUIOD, L. R., and HOLLENBERG, N. K.: Sodium intake and vascular smooth muscle responsiveness to norepinephrine and angiotensin in the rabbit. Circulation Res. *31:* 758–766 (1972).
33 MACDONALD, G. J.; BOYD, G. W., and PEART, W. S.: Effect of the angiotensin II blocker 1-sar-8-ala-angiotensin II on renal artery clip hypertension in the rat. Circulation Res. *37:* 640–646 (1975).

Dr. J. D. SWALES, University Department of Medicine, General Hospital, Gwendolen Road, *Leicester LE5 4PW* (England)

Discussion[1]

MAXWELL: Dr. FERRARIO, I was a little disturbed in your two-kidney model that at the end of 15 days you completely occluded the renal artery. In our experience and the experience of most people, when one occludes a renal artery you get a very transient hypertension and an atrophic kidney. I presume that there was some collateral circulation which developed in the first 15 days and this prevented renal atrophy. With graded renal artery constriction, one can get permanent hypertension without that.

While it does not seem appropriate to extrapolate from man to animals, nevertheless, the fact that the blood pressure did not decrease after analogue blockade in the chronic phase in your animals does not prove that renin or angiotensin was not still implicated in the blood pressure elevation, since it has been clearly shown in humans that, in chronic renovascular hypertension, one must have modest salt depletion to demonstrate the blood pressure lowering effect of saralasin. Did you try in the chronic phase to salt deplete your dogs and did they then not respond?

FERRARIO: We [FERRARIO et al.: in MILLIEZ and SAFAR Recent advances in hypertension, vol. 2, pp. 159–169, Boehringer Ingelheim, Reims 1975] and others [FEKETE et al.: Int. Urol. Nephrol. 2: 391–400, 1970] have shown that an occluded renal artery does not lead to renal insufficiency and kidney atrophy when collateral vessels are present. The technique used by us to produce a two-kidney model of renovascular hypertension in dogs was developed by Dr. MASAKI in my laboratory [Fed. Proc. Fed. Am. Socs exp. Biol. 35: 556, 1976]; it has allowed for producing sustained hypertension in dogs reliably and with regularity.

Second, yes we have tried sodium depletion. Plasma renin activity increases and blood pressure falls with administration of the angiotensin inhibitor. Dr. STREETEN has shown similar data in patients. If salt depletion is a prerequisite to demonstrate a blood pressure lowering effect of the angiotensin antagonist, then a salt factor, rather than renin may be primarily involved in the effect of angiotensin antago-

---

1 Discussion of the papers of FERRARIO et al. and SWALES.

nists on arterial pressure control. Our data suggest that the renin-angiotensin system does not participate *directly* in the control of arterial pressure during the chronic phase of renal hypertension. I believe that we are not alone in favouring other mechanisms to explain the maintenance of hypertension when increased renin secretion has abated to normal [McDonald et al.: Circulation Res. *37:* 640–646, 1975].

MAXWELL: In contrast to renovascular hypertension, in essential hypertension in man, with the same degree of salt depletion – causing a rise in PRA even though the base line is not necessarily high – saralasin has no blood pressure lowering effect.

LUMBERS: Dr. SWALES, reduction in pressor sensitivity to angiotensin II (AII) occurs in pregnancy and is also associated with a reduction in hand vascular sensitivity to AII. Are you postulating a blocking by arterially produced AII in the situations where you get a reduction in sensitivity to the exogenous angiotensin? Have you considered the role of angiotensinases in the reduction in pressor sensitivity?

SWALES: I think the data in pregnancy is a little contradictory. Here, we are concerned with acute changes in pressor sensitivity. No one would argue that in the long run renin is the only determinant of pressor sensitivity. I think that would be crazy, for instance, in view of the structural changes that occur in the vessel wall. There are other factors that are clearly important here. We are not arguing for endogenous production of AII within the blood vessel wall. In fact, all our evidence favours a renal origin. I agree that renal aminopeptidases could play a role in determining pressor sensitivity to angiotensin after nephrectomy. However, we get precisely the same changes after infusion of converting enzyme inhibitor, when presumably these aminopeptidases are not critically affected.

MACDONALD: Prof. SWALES, is one of the reasons that you got an enhanced pressor response in the presence of a reduced endogenous AII, such as following Squibb inhibitor or nephrectomy, the fact that, by reducing endogenous levels, you have moved yourself back along the dose-response curve (which it should be remembered is a log dose-response curve) so that the same arithmetical increase in plasma concentration will have a greater pressor increase associated with it than it would at a higher level of endogenous AII? I would also like to reinforce what other speakers have said, i.e. that there is a very close relationship between prevailing plasma renin levels (and, I believe, AII levels) and the subsequent depressor effect of angiotensin blockade, in renal artery clip hypertension.

This rather throws us back to thinking about circulating renin levels, and if we look for endocrine analogies it is very hard to find another endocrine disease where syndromes of oversecretion are not associated with a raised plasma level.

SWALES: I certainly agree with Dr. MACDONALD's view that there is a very close relationship between plasma angiotensin and the response to P113. This is not necessarily an argument either for or against our hypothesis since vascular renin, if it is of renal origin, will be proportional to circulating renin levels. I agree with the first statement. In fact, you can construct dose-response curves from the data of THURSTON and LARAGH and converting enzyme inhibitor does move the curve for low-salt pretreated rats very close to that of the high-salt pretreated rats.

BOYD: Dr. SWALES, crucial to this hypothesis is whether an angiotensin antibody will reduce the blood pressure rise with angiotensin infusions but not with renin. You showed us one experiment, but of course, we all know that different angiotensin antibodies may show very different effects. Therefore, is it the same antibody in both of these situations, and have you ever actually stripped the angiotensin off the antibody? I say this because antibodies may occasionally increase the angiotensin level unless care is taken to avoid angiotensin generation from continued renin enzymic activity in the antibody plasma *in vitro* before their injection.

SWALES: No, we have not done that experiment, and I think it would be a very good one. We have been using a constant antibody pool throughout the experiments, and this experiment was done with that antibody pool. I take your point: BING and NEILSON also looked at a similar situation, using angiotensin infusion. As far as I can remember their results, in a majority of cases antibody reduced the blood pressure to base line. A few antibodies did not; so clearly, as you say, it is critically dependent on the type of antibody used. However, the antibody tested against angiotensin-infusion hypertension was the one we used in the experiments reported here.

STREETEN: In connection with Dr. FERRARIO's comments on the relationship between renin and the response to saralasin, I would certainly agree that there is a statistically highly significant correlation between the two parameters when studied in large groups of subjects. However, this correlation breaks down very frequently in individual instances, where sodium retention, apparently by increasing the sensitivity of the receptors to AII, may result in hypertension, which is still AII-dependent, we believe. This angiotensin dependency persists in spite of the fact that the prevailing level of plasma renin activity and, I suspect, of AII, is technically 'normal', yet is probably better considered to be abnormal in the presence of the sodium retention in these patients. I would support Dr. MAXWELL's thesis, which our experience would confirm, that in essential hypertension, after raising the plasma renin activity by moderate salt depletion, one does not get a hypotensive response to saralasin, unless AII is the primary cause of the hypertension. Likewise, in women on contraceptive combinations who have coincidental essential hypertension, there is no hypotensive response to saralasin, even though the renin is very high.

HORVATH: Further to Dr. BOYD's questions of Prof. SWALES, we looked at different antibodies against AII, and in all the animals there was free-circulating AII. We could correlate the free AII levels with the K_m of the different antibodies and, in fact, there was an enormous amount of AII on each of the antibodies that could be leached off.

SWALES: Yes, I think that there is a difference between the *in vivo* and *in vitro* situation. I would certainly agree that antibodies do vary considerably though.

In STOKES and EDWARDS: Drugs Affecting the Renin-Angiotensin-Aldosterone System. Use of Angiotensin Inhibitors
Prog. biochem. Pharmacol., vol. 12, pp. 117–134 (Karger, Basel 1976)

The Use of Saralasin to Evaluate the Function of the Brain Renin-Angiotensin System[1]

IAN A. REID[2]

Department of Physiology, School of Medicine, University of California, San Francisco, Calif.

Contents

Introduction	117
Effects of Central Administration of Renin	119
Drinking	119
Blood Pressure	121
ADH Secretion	122
Effects of Central Administration of Substrate	124
Effects of Central Administration of Agents which Block the Renin-Angiotensin System	125
Drinking	126
Blood Pressure	127
Possible Role of Centrally Generated Angiotensin II in Hypertension	129
ADH Secretion	130
Concluding Remarks	130
Summary	131
References	132

Introduction

Since the original demonstration by BICKERTON and BUCKLEY [1] that angiotensin II can act directly on the central nervous system to increase

1   The previously unpublished work reported in this paper was supported by USPHS Grant AM06704 and the L. J. and Mary C. Skaggs Foundation.
2   Recipient of Research Career Development Award HL00104.

arterial blood pressure, it has become clear that this peptide has a variety of central actions. These include stimulation of drinking [2] and increased secretion of antidiuretic hormone (ADH) [3] and adrenocorticotropic hormone (ACTH) [4]. In addition, there is evidence that angiotensin may inhibit norepinephrine uptake by central adrenergic neurones [5] and increase the release of acetylcholine at central cholinergic nerve terminals [6].

Information concerning the sites of action of angiotensin on the central nervous system is also becoming available. The pressor effects appear to result from an action at two sites – the area postrema [7] and the subnucleus medialis [8]. There is evidence that the stimulation of ADH secretion by angiotensin results from an action on the supraoptic nucleus [9, 10], but a direct action on the posterior pituitary has also been suggested [11]. The subfornical organ appears to contain the receptors for the dipsogenic action of angiotensin [12], but other sites have also been implicated [2]. Angiotensin may act directly on the anterior pituitary to increase the secretion of ACTH [4].

Despite this progress, it is not clear if these effects of angiotensin represent physiological actions of the peptide. In general, the doses of angiotensin used to elicit the effects have been very large and would have produced concentrations of angiotensin well above the physiological range. Another problem is that angiotensin is a polar compound and does not appear to cross the blood-brain barrier. This particular problem has been partially solved by the demonstration that the receptors for at least some of the central actions of angiotensin are in areas that are 'outside the blood-brain barrier' and are therefore accessible to circulating angiotensin. These areas include the subfornical organ, the area postrema and the pituitary. However, there is no evidence that the subnucleus medialis or the supraoptic nucleus are 'outside the blood-brain barrier' and it is therefore difficult to see how circulating angiotensin could influence these structures.

A possibility that has aroused considerable interest in recent years is that angiotensin is actually synthesized within the brain and would therefore be available to interact with angiotensin receptors that are 'within the blood-brain barrier'. This suggestion originated with the demonstration of renin-like activity in extracts of dog and rat [13, 14] and more recently human [15] brain. This material resembles renal renin in that its pH optimum is low but differs from renal renin in that it has little or no activity at pH 7.5 [13, 15]. Recent studies in this laboratory suggest that the renin activity in brain may be due to the lysosomal acidic protease cathepsin D

[16]. Renin substrate (angiotensinogen) is also present in the central nervous system both in brain tissue [13, 17] and cerebrospinal fluid [18]. Enzymes with the ability to convert angiotensin I to angiotensin II and to degrade angiotensin II are present in brain tissue [19, 20], but not in cerebrospinal fluid [21; REID et al., unpublished observations]. There have been reports that substances resembling angiotensin I and II are present in brain tissue and/or cerebrospinal fluid [13, 14, 22], but there are discrepancies with regard to concentration and distribution. There is little information concerning the biochemical, immunological and pharmacological properties of these substances and further characterization is required.

Do the components of the renin-angiotensin system in the brain interact to form a functional system and if so, does the system play a physiological role? One approach to answering these questions is to determine if renin is active when administered centrally. Since all the known actions of renin are mediated via angiotensin II, the finding that renin is active when administered centrally would indicate that there is an interaction between the injected renin, brain angiotensinogen and converting enzyme that results in the formation of biologically active amounts of angiotensin II. Similarly, the finding that central administration of angiotensinogen produces physiological effects would constitute evidence for the existence of renin activity in brain under physiological conditions.

Another approach is to study the effects of central administration of inhibitors of the renin-angiotensin system. If variables such as blood pressure, water intake and ADH secretion are normally influenced by a brain renin-angiotensin system, it should be possible to alter them by blocking the system with renin inhibitors (e.g. pepstatin), converting enzyme inhibitors (e.g. SQ 20881) or angiotensin antagonists (e.g. saralasin).

The purpose of this chapter is to describe experiments which have recently been performed in our laboratory along the lines discussed above, and to review the results of similar investigations performed in other laboratories.

Effects of Central Administration of Renin

Drinking

The effect on drinking of central administration of renin was studied in eight mongrel dogs [18]. Hog renin (0.1 Goldblatt units) was injected directly into the ventricular system through a cannula implanted chronically in the third ventricle. This treatment stimulated drinking in each of the dogs; the response had a latency of approximately 2 min and the mean volume of water drunk in the 15-min period following the injection was 485 ± 84 ml. The same dose of renin did not stimulate drinking when administered intravenously. To determine if the dipsogenic action of intraventricular renin was mediated via the formation of angiotensin II, the effect of saralasin on the response was studied. The saralasin was administered in two 10-μg doses; the first was injected 2 min before the renin while the second was injected with the renin. Administered in this fashion, saralasin abolished the drinking response in seven of the dogs and markedly attenuated it in the eighth; the mean volume of water drunk was reduced to 8 ± 6 ml.

These results are in good agreement with those obtained in other species. EPSTEIN et al. [23] observed that the drinking response to intracranial renin in rats could be blocked by pepstatin, SQ 20881 and saralasin. Similar results were obtained in the cat by COOLING and DAY [24]. Thus, it is clear that renin is an effective dipsogen when administered centrally and that this action is mediated via the formation of angiotensin II.

Blood Pressure

The effect of intraventricular administration of renin on arterial pressure was studied in 13 dogs anaesthetized with sodium pentobarbital [18]. Both hog and dog renin were administered, in doses ranging from 0.05 to 0.25 Goldblatt units. For comparison, the effect of intraventricular angiotensin II, in doses ranging from 0.1 to 1.0 μg, was studied. The same doses of renin and angiotensin II were also administered intravenously. The results of a representative study are shown in figure 1. Intraventricular renin consistently increased arterial pressure. The response had a latency of 0.5–1.5 min and averaged 16 mm Hg. The increase was very prolonged, lasting from 30 min to more than 3 h. Intraventricular angiotensin II produced a similar increase in blood pressure but the duration of the response was shorter than that produced by renin. Intravenous administration of the same doses of renin and angiotensin produced transient pressor responses.

The effect of saralasin on the blood pressure response to intraventricular renin was studied in five dogs. The antagonist was administered in the same way as in the drinking experiments. The results are summarized in figure 2. In the absence of saralasin, renin increased arterial pressure from 89 ± 5 to 107 ± 5 mm Hg ($p<0.01$). When the renin was preceded by saralasin, the blood pressure response was abolished (85 ± 5 to

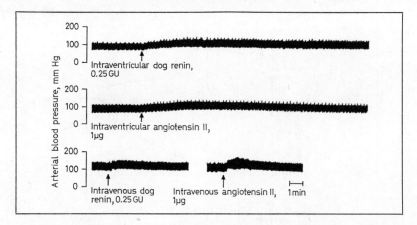

Fig. 1. Comparison of the effects of intraventricular and intravenous renin and angiotensin II on arterial blood pressure in an anesthetized dog. Reproduced from REID and RAMSAY [18].

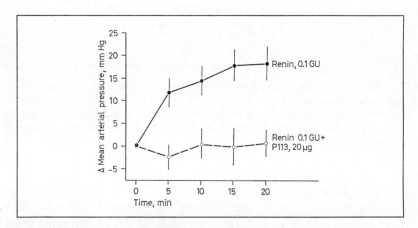

Fig. 2. The effect of intraventricular saralasin on the blood pressure response to intraventricular renin. Vertical bars represent the mean ± SEM of observations in 5 dogs. Reproduced from REID and RAMSAY [18].

86 ± 7 mm Hg). Intraventricular saralasin did not reduce the blood pressure response to intravenous angiotensin II.

Taken together, these results indicate that renin increases arterial pressure when administered centrally and that this effect is mediated via the formation of angiotensin II within the central nervous system.

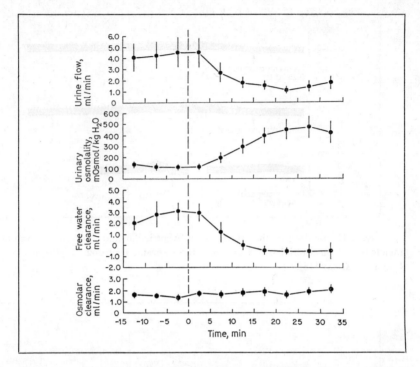

Fig. 3. The effects of intraventricular renin on urine flow, urinary osmolality, free water clearance and osmolar clearance in anesthetized dogs undergoing a water diuresis. Renin was injected at time 0 in a dose of 0.1 Goldblatt units. Each point represents the mean ± SEM of observations in 5 dogs. Reproduced from MALAYAN and REID [25].

ADH Secretion

The effect of centrally administered renin on ADH secretion was initially investigated in pentobarbital-anesthetized dogs undergoing a water diuresis [25]. In these experiments, ADH was not measured directly but instead, changes in urine flow, free water clearance and urinary osmolality were used as an index of changes in ADH secretion. Hog renin was injected into the third ventricle in a dose of 0.1 Goldblatt units. The results are summarized in figure 3. Intraventricular renin produced a prompt decrease in urine flow from 4.5 ± 1.4 to 1.1 ± 0.2 ml/min ($p<0.04$), in association with a decrease in free water clearance from 3.0 ± 1.3 to −0.6 ± 0.3 ml/min ($p<0.03$) and an increase in urinary osmolality from 106 ± 20 to 449 ± 90 mOsmol/kg H_2O ($p<0.03$). There was no signifi-

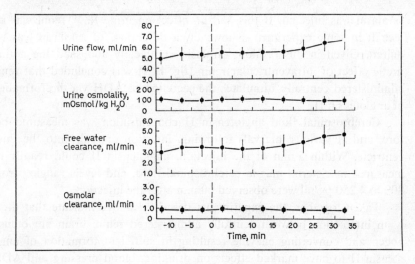

Fig. 4. The effects of intraventricular renin on urine flow, urinary osmolality, free water clearance and osmolar clearance in dogs pretreated with saralasin (see legend for fig. 3). Data from MALAYAN and REID [25].

cant change in osmolar clearance. This antidiuretic effect was accompanied by an increase in urinary sodium excretion from 59 ± 17 to 137 ± 32 µEq/min ($p<0.03$); potassium excretion did not change. Each of these changes could be prevented by hypophysectomy.

Although these data strongly suggest that intraventricular renin stimulates the secretion of ADH, additional experiments were performed in which plasma ADH concentration was measured by radioimmunoassay [3]. In a group of nine pentobarbital-anesthetized dogs, intraventricular renin (0.1 Goldblatt units) caused a prompt increase in plasma ADH concentration from 11.7 ± 2.3 to 17.5 ± 4.3 pg/ml at 1.5 min ($p<0.002$) and to 23.8 ± 5.6 pg/ml at 5 min ($p<0.05$). Plasma ADH concentration was still elevated 30 min after the injection. The same dose of renin did not affect plasma ADH concentration when administered intravenously.

Having demonstrated that changes in free water clearance and urinary osmolality are a valid index of changes in ADH secretion, the effect of saralasin on the antidiuretic effect of intraventricular renin was studied. Saralasin was initially injected into the third ventricle in a dose of 10 µg. It was found that the antagonist possessed some agonist activity and produced a transient antidiuresis resembling that produced by intraventricular renin. When the diuresis had reversed, a second 10 µg dose of saralasin was injected. If this had no effect on urine flow, renin was injected; in some experiments, however, a third dose of saralasin was required. Given in this fashion, saralasin completely abolished the antidiuretic effect of intraventricular renin (fig. 4). It was concluded that renin administered centrally stimulates the secretion of ADH via the formation of angiotensin II.

Cerebrospinal fluid angiotensin II concentration was measured before, and at various intervals following, injection of renin into the third ventricle. Within 5 min of the injection, angiotensin II could readily be measured in cisterna magna cerebrospinal fluid, and levels ranging from 809 to 4,250 pg/ml were observed 60 min after the injection.

Taken together, the experiments described above indicate that there is an interaction between centrally administered renin, brain angiotensinogen and converting enzyme resulting in sufficient formation of angiotensin II to have marked effects on drinking, blood pressure and ADH secretion.

Effects of Central Administration of Substrate

It is probably a reasonable assumption that saturating levels of renin substrate are not present in the brain since the concentration of angiotensinogen in cerebrospinal fluid in dogs is only one fifth of the plasma angiotensinogen concentration [18]; the angiotensinogen concentration in brain tissue is also much less than in plasma [13, 17]. Thus, an approach to the question of whether endogenous brain renin activity can produce the same effects as central injection of renin is to determine if central administration of renin substrate also results in the formation of angiotensin II in the brain.

Such studies have utilized the synthetic tetradecapeptide substrate (the N-terminal sequence of human and hog angiotensinogen). EPSTEIN et al. [23] observed that intracranial injection of the tetradecapeptide stimulated drinking in rats. This response was reduced slightly by pepstatin and markedly attenuated by SQ 20881 or saralasin. It was therefore concluded that the drinking response to the tetradecapeptide was mediated via the local generation of angiotensin II.

In preliminary experiments, we have found intraventricular tetradecapeptide to cause a prompt increase in blood pressure (fig. 5) which appeared to be shorter lived than that produced by intraventricular renin or

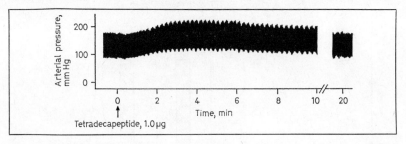

Fig. 5. Effect of intraventricular tetradecapeptide renin substrate on blood pressure in an anesthetized dog.

angiotensin II (fig. 1). The effects of inhibitors of the renin-angiotensin system on this response have not yet been investigated.

The results of these studies should be interpreted with caution as there are marked differences between the properties of the tetradecapeptide and the naturally occurring substrate. For example, the affinity of brain renin activity for the tetradecapeptide is much greater than for plasma angiotensinogen [13, 15]. Indeed, DAUL et al. [15] observed that human brain 'renin', which readily reacted with the tetradecapeptide, had negligible activity against human plasma angiotensinogen. In addition to this difference in affinity, there is a difference between the pH optimum for the renin-tetradecapeptide reaction and the renin-plasma angiotensinogen reaction [26]. Finally, it is possible that the central dipsogenic and pressor effects of the synthetic substrate are not due to its being cleaved by renin. For example, it has been reported that converting enzyme can split angiotensin II from the tetradecapeptide [27]. In addition, SKEGGS et al. [28] reported that the tetradecapeptide itself possesses some biological activity and this could conceivably be responsible, at least in part, for the central effects described above. It therefore seems premature to interpret the central actions of the synthetic substrate as being evidence for an action of brain renin, and additional experiments utilizing the naturally occurring substrate are required.

Effects of Central Administration of Agents which Block the Renin-Angiotensin System

Injection of inhibitors of the renin-angiotensin system into the central nervous system should reveal any actions of angiotensin II formed within

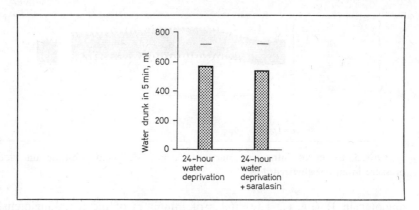

Fig. 6. Effect of intraventricular saralasin on the drinking response to 24-hour water deprivation in 5 dogs. Data from RAMSAY and REID [29].

the brain, although they may also block some of the central effects of blood-borne angiotensin II and this must be taken into account in the interpretation of such experiments.

Drinking
The effect of central administration of saralasin on the drinking response to water deprivation has been investigated in this laboratory [29]. Five dogs were deprived of water for 24 h; at the end of this period, they were presented with water and the volume consumed over the next 5 min monitored. On a separate occasion, the same procedure was repeated except that saralasin was injected into the third ventricle immediately prior to presentation of water. The saralasin was administered in two 10-μg doses separated by 2 min (this dose of saralasin blocks the drinking response to intraventricular renin, see above). The volume of water consumed was not affected by saralasin (fig. 6). Following 24 h water deprivation in the control experiments, the dogs drank 570 ± 146 ml; following treatment with saralasin, they drank 537 ± 184 ml. Thus, no role for angiotensin, generated either centrally or peripherally, in the drinking response to water deprivation was demonstrated.

The responses to other stimuli to drinking are also unaffected by centrally administered blockers of the renin-angiotensin system. SUMMY-LONG and SEVERS [30] reported that intraventricular administration of saralasin or SQ 20881 did not reduce the drinking produced in rats by relative cellular dehydration (hypertonic NaCl injection) or hypovolaemia

Table I. Effect of intraventricular saralasin on blood pressure and plasma ADH concentration in anesthetized dogs [MALAYAN et al., unpublished observations]

	Time, min							
	−5	0	1.5	3	5	10	15	30	60
Mean Arterial Pressure mm Hg	106 ±6	108 ±7	112 ±8	105 ±8	109 ±5	108 ±5	105 ±6	102 ±8	100 ±7
Plasma ADH Concentration pg/ml	12.2 ±4.2	13.6 ±5.7	14.4 ±3.7	15.1 ±7.6	19.8 ±8.7	20.7 ±7.7	21.8 ±9.0	23.5 ±8.7	16.2 ±4.3

Saralasin was injected at time 0 in a dose of 20 µg. Values represent the mean ± SEM of observations in 6 animals.

(hyperosmotic polyethylene glycol injection). Similarly, LEHR et al. [31] observed that intraventricular administration of SQ 20881 in rats did not modify the drinking response to either isoproterenol administration or caval ligation. None of these experiments provides evidence for the participation of a brain renin-angiotensin system in the regulation of drinking.

Blood Pressure

During recent years, evidence has been accumulating that the renal renin-angiotensin system participates in the regulation of blood pressure, not only through the direct action of angiotensin on vascular smooth muscle and aldosterone secretion, but also through its effects on the central nervous system [7, 32]. It has been suggested that the brain renin-angiotensin system might also influence central mechanisms of blood pressure regulation [33].

As a first approach to testing this possibility, the effect of central administration of saralasin on arterial pressure in anesthetized dogs was studied. In the first group of experiments, the antagonist was injected into the third ventricle in a single dose of 20 µg. This dose blocks the pressor effect of intraventricular renin (fig. 2). The results are summarized in table I. Intraventricular saralasin had no significant effect on blood pressure.

In a second group of experiments, the effect of intraventricular saralasin on blood pressure in sodium-deficient dogs was studied. Sodium deficiency was produced by feeding the dogs a low sodium diet for approxi-

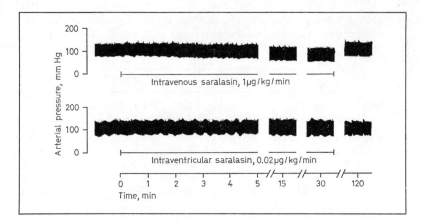

Fig. 7. Comparison of the effects of intravenous and intraventricular saralasin on blood pressure in a sodium-deficient dog.

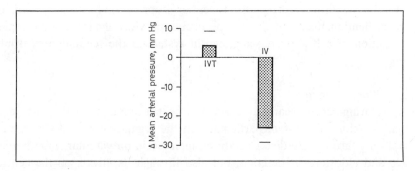

Fig. 8. Comparison of the effects of intravenous (IV) and intraventricular (IVT) saralasin on blood pressure in 4 sodium-deficient dogs.

mately one week; in addition, furosemide (20 mg i.m.) was administered for the first two days of this period. The animals were anaesthetized with sodium pentobarbital and saralasin was infused either intravenously or directly into the third ventricle for 15–30 min. The results are summarized in figures 7 and 8. Intraventricular infusion of saralasin at a rate of 0.02 μg/kg/min did not produce any significant changes in blood pressure. This dose of saralasin prevented the pressor response to intraventricular injection of 200 ng angiotensin II, indicating effective blockade of central angiotensin receptors. In marked contrast, intravenous infusion of saralasin in a dose of 1 µg/kg/min consistently decreased blood pressure. These experiments thus failed to reveal any evidence for the participation of a brain renin-angiotensin system in the regulation of blood pressure in either normal or sodium-depleted dogs.

Possible Role of Centrally Generated Angiotensin II in Hypertension

FINKIELMAN et al. [34] reported that a pressor peptide pharmacologically similar to angiotensin I was present in the cerebrospinal fluid of normotensive and hypertensive patients. Furthermore, a close correlation was found between the concentration of the peptide and the blood pressure of a group of essential hypertensive patients. It was felt that this material was synthesized within the central nervous system since angiotensin does not cross the blood-brain barrier. This report thus provided the first suggestive evidence for a role of centrally generated angiotensin in hypertension.

The possible role of central angiotensin in the control of blood pressure in spontaneously hypertensive rats has been studied by a number of investigators. GANTEN et al. [33] reported that the concentration of angiotensin II in the cerebrospinal fluid of these rats is higher than in normotensive controls. They observed that intraventricular administration of saralasin decreased blood pressure in the hypertensive, but not the normotensive rats; this effect could also be produced by prolonged perfusion of the ventricular system with artificial cerebrospinal fluid [35]. PHILLIPS et al. [36] confirmed that intraventricular administration of saralasin lowers blood pressure in spontaneously hypertensive rats, and SWEET et al. [37] showed that another angiotensin II antagonist. Sar^1-Ile^8-angiotensin II is also effective. These experiments indicate that centrally generated angiotensin may participate in the control of blood pressure in the spontaneously hypertensive rat. On the other hand, ELGHOZI et al. [38] reported that neither Sar^1-Ile^8-angiotensin II nor Sar^1-Thr^8-angiotensin II decreased blood pressure in spontaneously hypertensive rats. The reason for this discrepancy is not clear.

Intraventricular administration of Sar^1-Ile^8-angiotensin II has also been reported to lower blood pressure in rats with malignant hypertension [37]. However, central administration of pepstatin or a converting enzyme inhibitor failed to lower blood pressure, suggesting that the effectiveness of the angiotensin antagonist was due to blockade of peripheral, rather than central angiotensin II. It appears that this is the case since bilateral nephrectomy also lowered blood pressure in these rats and abolished the blood pressure lowering action of the antagonist [SWEET, personal commun.].

In summary, there is no evidence for the participation of a central renin-angiotensin system in blood pressure regulation in normal, sodium-deficient or renal hypertensive animals. On the other hand, centrally generated angiotensin appears to play a role in spontaneously hypertensive rats.

ADH Secretion

The supraoptic nucleus contains specific receptors for angiotensin II [9, 10], but the physiological significance of these receptors is not clear. Most of the available data suggest that the supraoptic nucleus is 'inside the blood-brain barrier' and therefore inaccessible to blood borne angiotensin II. For example, NICOLL and BARKER [9] reported that supraoptic neurosecretory cells were excited by direct application of angiotensin II, but were unaffected by angiotensin administered via the carotid artery. In addition, although it is well established that centrally administered angiotensin II stimulates ADH secretion [3, 39], it is not clear if systemically administered angiotensin can produce the same effect [40, 41].

A possible explanation for these findings is that the supraoptic neurosecretory cells are normally influenced by centrally, rather than peripherally generated angiotensin II. If this is so, it should be possible to decrease ADH secretion by central administration of antagonists of the renin-angiotensin system. The effect of saralasin on ADH secretion in dogs was therefore studied. The antagonist was injected into the third ventricle in a single dose of 20 μg – this dose blocks the stimulation of ADH secretion by intraventricular renin (fig. 4). Following intraventricular saralasin, plasma ADH concentration either did not change or increased; in no animals did plasma ADH decrease. The mean values are shown in table I – none of these values is significantly different from the control values. Thus, these preliminary experiments do not provide evidence for a role of centrally generated angiotensin in the control of ADH secretion.

Concluding Remarks

All the components required for the generation of angiotensin II are present in the central nervous system, but it is not clear if these interact to form a functional system analogous to the peripheral renin-angiotensin system. The finding that renin increases water intake, blood pressure and ADH secretion when administered centrally and that these effects can be blocked by central saralasin, indicates an interaction between the injected renin, brain angiotensinogen and converting enzyme, resulting in the formation of physiologically active amounts of angiotensin II. However, it has not been satisfactorily demonstrated that endogenous brain renin activity can also produce these effects. Indeed, there is evidence that the renin activity measured in brain extracts is due to the lysosomal acidic protease cathepsin D [16]; this enzyme has little or no activity above pH 6.0.

Experiments utilizing central administration of blockers of the renin-angiotensin system have in general failed to provide evidence for the existence of a functional brain renin-angiotensin system. When injected into the cerebral ventricles, these agents do not appear to decrease water intake, ADH secretion or, with the exception of the spontaneously hypertensive rat, arterial blood pressure. Additional investigation is clearly required to further evaluate the possible existence and functional significance of the proposed brain renin-angiotensin system.

Summary

The demonstration that the components required for the generation of angiotensin II are present in the brain has led to the proposal that there is a brain renin-angiotensin system. To test this hypothesis, experiments were performed to determine if biologically active amounts of angiotensin II are formed when renin is injected into the cerebral ventricles. The effects of central administration of agents known to block the peripheral renin-angiotensin system were also investigated. It was shown that intraventricular renin increased water intake, blood pressure and ADH secretion and that these effects were blocked by saralasin. These findings indicated an interaction between injected renin, brain angiotensinogen and converting enzyme, resulting in the formation of angiotensin II in physiologically active concentrations. However, these experiments did not demonstrate a role for endogenous brain renin activity. Central administration of saralasin in normal animals did not decrease water intake, blood pressure or ADH secretion. These studies thus failed to demonstrate a physiological role for the proposed brain renin-angiotensin system in controlling water balance and blood pressure.

References

1 BICKERTON, R. K. and BUCKLEY, J. P.: Evidence for a central mechanism in angiotensin-induced hypertension. Proc. Soc. exp. Biol. Med. *106:* 834–836 (1961).
2 SEVERS, W. B. and SUMMY-LONG, J.: The role of angiotensin in thirst. Life Sci. *17:* 1513–1526 (1975).
3 KEIL, L. C.; SUMMY-LONG, J., and SEVERS, W. B.: Release of vasopressin by angiotensin II. Endocrinology *96:* 1063–1065 (1975).
4 MARAN, J. W. and YATES, F. E.: Locus of ACTH-releasing action of angiotensin II. Endocrinology *94:* A118 (1974).
5 PALAIC, D. and KHAIRALLAH, P. A.: Effect of angiotensin on uptake and release of norepinephrine by brain. Biochem. Pharmacol. *16:* 2291–2298 (1967).
6 ELIE, R. and PANISSET, J.-C.: Effect of angiotensin and atropine on the spontaneous release of acetylcholine from cat cerebral cortex. Brain Res. *17:* 297–305 (1970).
7 FERRARIO, C. M.; GILDENBERG, P. L., and MCCUBBIN, J. W.: Cardiovascular effects of angiotensin mediated by the central nervous system. Circulation Res. *30:* 257–262 (1972).
8 DEUBEN, R. R. and BUCKLEY, J. P.: Identification of a central site of action of angiotensin II. J. Pharmac. exp. Ther. *175:* 139–146 (1970).
9 NICOLL, R. A. and BARKER, J. L.: Excitation of supraoptic neurosecretory cells by angiotensin II. Nature new Biol. *233:* 172–174 (1971).
10 SAKAI, K. K.; MARKS, B. H.; GEORGE, J., and KOESTNER, A.: Specific angiotensin II receptors in organ-cultured canine supra-optic nucleus cells. Life Sci. *14:* 1337–1344 (1974).
11 GAGNON, D. J.; COUSINEAU, D., and BOUCHER, P. J.: Release of vasopressin by angiotensin II and prostaglandin E_2 from the rat neuro-hypophysis *in vitro*. Life Sci. *12:* 487–497 (1973).
12 SIMPSON, J. B. and ROUTTENBERG, A.: Subfornical organ: site of drinking elicitation by angiotensin II. Science *181:* 1172–1175 (1973).
13 GANTEN, D.; MARQUEZ-JULIO, A.; GRANGER, P.; HAYDUK, K.; KARSUNKY, K. P.; BOUCHER, R., and GENEST, J.: Renin in dog brain. Am. J. Physiol. *221:* 1733–1737 (1971).
14 FISCHER-FERRARO, C.; NAHMOD, V. E.; GOLDSTEIN, D. J., and FINKIELMAN, S.: Angiotensin and renin in rat and dog brain. J. exp. Med. *133:* 353–361 (1971).
15 DAUL, C. B.; HEATH, R. G., and GAREY, R. E.: Angiotensin-forming enzyme in human brain. Neuropharmacology *14:* 75–80 (1975).
16 DAY, R. P. and REID, I. A.: Renin activity in dog brain: enzymological similarity to cathepsin D. Endocrinology *99:* 93–100 (1976).
17 PRINTZ, M. P. and LEWICKI, J.: Renin substrate in the central nervous system: potential significance to central regulatory mechanisms; in BUCKLEY and FERRARIO The central actions of angiotensin and related peptides (Pergamon Press, Oxford, in press).
18 REID, I. A. and RAMSAY, D. J.: The effects of intracerebroventricular adminis-

tration of renin on drinking and blood pressure. Endocrinology 97: 536–542 (1975).
19 YANG, H-Y. T. and NEFF, N. H.: Distribution and properties of angiotensin converting enzyme of rat brain. J. Neurochem. 19: 2443–2450 (1972).
20 GOLDSTEIN, D. J.; DIAZ, A.; FINKIELMAN, S.; NAHMOD, V. E., and FISCHER-FERRARO, C.: Angiotensinase activity in rat and dog brain. J. Neurochem. 19: 2451–2452 (1972).
21 GANTEN, D.; GANTEN, U.; SCHELLING, P.; BOUCHER, R., and GENEST, J.: The renin and iso-renin-angiotensin systems in rats with experimental pituitary tumors. Proc. Soc. exp. Biol. Med. 148: 568–572 (1975).
22 SAAD, W. A.; EPSTEIN, A. N.; SIMPSON, J. B., and CAMARGO, L. A.: Brain and blood-borne angiotensin II in the control of thirst. Neurosci. Abstr. 1: 470 (1975).
23 EPSTEIN, A. N.; FITZSIMMONS, J. T., and JOHNSON, A. K.: Peptide antagonists of the renin-angiotensin system and the elucidation of the receptors for angiotensin-induced drinking. J. Physiol., Lond. 238: 34P–35P (1974).
24 COOLING, M. J. and DAY, M. D.: Inhibition of renin-angiotensin induced drinking in the cat by enzyme inhibitors and by analogue antagonists of angiotensin II. Clin. exp. Pharmacol. Physiol. 1: 389–396 (1974).
25 MALAYAN, S. A. and REID, I. A.: Antidiuresis produced by injection of renin into the third cerebral ventricle of the dog. Endocrinology 98: 329–335 (1976).
26 FAVRE, L.; ROUSSEL-DERUYCK, R., and VALLOTTON, M. B.: Influence of pH on human renin activity with different substrates: role of substrate denaturation. Biochim. biophys. Acta 302: 102–109 (1973).
27 DORER, F. E.; KAHN, J. R.; LENTZ, K. E.; LEVINE, M., and SKEGGS, L. T.: Formation of angiotensin II from tetradecapeptide renin substrate by angiotensin-converting enzyme. Biochem. Pharmacol. 24: 1137–1139 (1975).
28 SKEGGS, L. T.; LENTZ, K. E.; KAHN, J. R.; DORER, F. E., and LEVINE, M.: Pseudorenin – a new angiotensin-forming enzyme. Circulation Res. 25: 451–462 (1969).
29 RAMSAY, D. J. and REID, I. A.: Some central mechanisms of thirst in the dog. J. Physiol., Lond. 253: 517–525 (1975).
30 SUMMY-LONG, J. and SEVERS, W. B.: Angiotensin and thirst: studies with a converting enzyme inhibitor and a receptor antagonist. Life Sci. 15: 569–582 (1974).
31 LEHR, D.; GOLDMAN, H. W., and CASNER, P.: Renin-angiotensin role in thirst: paradoxical enhancement of drinking by angiotensin converting enzyme inhibitor. Science 182: 1031–1033 (1973).
32 SEVERS, W. B. and DANIELS-SEVERS, A. E.: Effects of angiotensin on the central nervous system. Pharmacol. Rev. 25: 415–449 (1973).
33 GANTEN, D.; HUTCHINSON, J. S., and SCHELLING, P.: The intrinsic brain iso-renin-angiotensin system in the rat: its possible role in central mechanisms of blood pressure regulation. Clin. Sci. mol. Med. 48: 265S–268S (1975).
34 FINKIELMAN, S.; FISCHER-FERRARO, C.; DIAZ, A.; GOLDSTEIN, D. J., and NAHMOD, V. E.: A pressor substance in the cerebrospinal fluid of normotensive and hypertensive patients. Proc. natn. Acad. Sci. USA 69: 3341–3344 (1972).
35 GANTEN, D.; HUTCHINSON, J. S.; SCHELLING, P.; GANTEN, U., and FISCHER, H.:

The iso-renin angiotensin systems in extrarenal tissue. Clin. exp. Pharmacol. Physiol. *3:* 103–126 (1976).

36 PHILLIPS, M. I.; PHIPPS, J.; HOFFMAN, W., and LEAVITT, M.: Reduction of blood pressure by intracranial injection of angiotensin blocker (P113) in spontaneously hypertensive rats. Physiologist *18:* 350 (1975).

37 SWEET, C. S.; COLUMBO, J. M., and GAUL, S. L.: Inhibitors of the renin-angiotensin system in the malignant hypertensive rat: comparative antihypertensive effects of central vs. peripheral administration; in BUCKLEY and FERRARIO The central actions of angiotensin and related peptides (Pergamon Press, Oxford, in press).

38 ELGHOZI, J. L.; ALTMAN, J.; DEVYNCK, M. A.; LIARD, J. F.; GRUNFELD, J. P., and MEYER, P.: Lack of hypotensive effect of central injection of angiotensin inhibitors in spontaneously hypertensive and normotensive rats; in BUCKLEY and FERRARIO The central actions of angiotensin and related peptides (Pergamon Press, Oxford, in press).

39 SEVERS, W. B.; SUMMY-LONG, J.; TAYLOR, J. S., and CONNOR, J. D.: A central effect of angiotensin: release of pituitary pressor material. J. Pharmac. exp. Ther. *174:* 27–34 (1970).

40 CLAYBAUGH, J. R.; SHARE, L., and SHIMIZU, K.: The inability of infusions of angiotensin to elevate plasma vasopressin concentration in the anesthetized dog. Endocrinology *90:* 1647–1652 (1972).

41 BONJOUR, J. P. and MALVIN, R. L.: Strimulation of ADH release by the renin-angiotensin system. Am. J. Physiol. *218:* 1555–1559 (1970).

Dr. I. A. REID, Department of Physiology, School of Medicine, University of California, *San Francisco CA 94143* (USA)

Competitive Inhibitors of Renin[1]
A Review

KNUD POULSEN, JAMES BURTON and EDGAR HABER

The University Institute for Experimental Medicine, Copenhagen; Cardiac Unit, Massachusetts General Hospital, and Department of Medicine, Harvard Medical School, Boston, Mass.

Contents

Introduction 135
Results and Discussion 136
Summary 140
References 140

Introduction

The basis for our investigation was the work of SKEGGS et al. [1], who determined the amino acid sequence around the cleavage site of renin substrate and also determined the smallest peptide with the native amino acid sequence which could be cleaved by renin at a reasonable rate. This peptide was an octapeptide and had the sequence His-Pro-Phe-His-Leu-Leu-Val-Tyr. An even shorter peptide Leu-Leu-Val-tyrosinol was synthesized by KOKUBO et al. [2] and, together with a series of related peptides, was shown to be a competitive although weak inhibitor for renin. We first resynthesized the octapeptide and confirmed that it was a substrate for renin, and could be cleaved at the Leu-Leu bond [3]. This peptide, with the native amino acid sequence, was not only cleaved by renin,

1 This work was supported by the Danish Heart Association, the Danish Medical Research Council, in part by SCOR HL-14150 from the National Institutes of Health, Nas 9-10981, and USPHS HL 06664.

but also acted as a competitive substrate when human renin reacted with the tetradecapeptide. The tetradecapeptide, Asp-Arg-Val-Tyr-Ile-His-Pro-Phe-His-Leu-Leu-Val-Tyr-Ser, is a larger peptide with the native sequence, which when cleaved by renin, forms angiotensin I Asp-Arg-Val-Tyr-Ile-His-Pro-Phe-His-Leu [1]. The octapeptide acted therefore as a competitive inhibitor since its cleavage product was too short to be angiotensin I or to have immunological similarities with angiotensin I. However, the octapeptide was cleaved by renin. A much more efficient inhibitor would be a peptide with affinity for the enzyme site of renin but which was not cleaved. Substitutions by D-amino acids and L-amino acids of somewhat differing structure were thus performed in the octapeptide. These substitutions were selected because of prior demonstrations that such changes in substrates for other enzymes resulted in competitive inhibitors. This has been shown for carboxypeptidase A [4], chymotrypsin [5] and papain [6].

Results and Discussion

Some of the peptides initially synthesized [3] are listed in figure 1. It is seen that substitutions can be made throughout the octapeptide without decreasing the affinity. On the contrary, the affinity was increased in most cases, as judged by the decrease in K_i. Even the two leucine molecules at the cleavage site could be substituted; if they were substituted with D-amino acids, the molecules could no longer be cleaved by renin. Of the peptides a D-leucine substitution in position 6, D-Leu^6-octapeptide, was the best competitive inhibitor, since it could not be cleaved by renin and had a K_i of 3 μM, which corresponds to an affinity one order of magnitude higher than that of the native octapeptide. The study of PARIKH and CUATRECASAS [7] confirms that replacement with D-amino acids at the site of cleavage in the native sequence of the tetradecapeptide yields effective competitive inhibitors of human renin. The D-Leu^6-octapeptide was, like the other inhibitors, a true competitive inhibitor, judged by weighted Lineweaver-Burk plots. The use of the inhibitors, however, was restricted to *in vitro* studies because they were not very soluble at physiological pH.

A study was then undertaken to increase the solubility [8]. The best methods involved a prolongation of the peptide chain with from one to five proline molecules added to the N-terminal of the peptide. This greatly increased the solubility so that the inhibitory capacity of the peptides

1	2	3	4	5	6	7	8	K_i μM	Cleavage %
His	Pro	Phe	His	Leu	Leu	Val	Tyr	39	97
D-His								6	30
							Phe	3	100
				D-Leu				25	0
					D-Leu			3	0
				ileu				7	30
					ileu			>25	20
							tyrosinol	1,020	
Tetradecapeptide $K_m=143$									

Fig. 1. Inhibitors and substrates of human renin, pH 5.5.

	1	2	3	4	5	6	7	8	K_i μM	Solubility μM
	His	Pro	Phe	His	Leu	Leu	Val	Tyr	–	161
Pro									42	324
						D-Leu			–	137
Pro						D-Leu			–	41
Pro-Pro-Pro						D-Leu			–	365
Pro						Tyr			12	303
Pro						Phe			4	412
Pro					Phe	Phe			1	100

Fig. 2. Human renin inhibitors, pH 7.5.

could be demonstrated. In figure 2, the solubilizing effect of proline is demonstrated for the octapeptide. Proline addition also increased the solubility of D-Leu[6]-octapeptide but, to our surprise, this peptide, which was the best inhibitor at pH 5.5, was completely without effect at physiological pH. Fortunately, other substitutions showed inhibitory properties at physiological pH. This is especially true when hydrophobic residues such as tyrosine and phenylalanine are substituted for leucine in position 6. The best inhibitor so far is a double substitution with a phenylalanine substituted for each of the two leucines and solubilised by an N-terminal proline. This peptide is a true and effective competitive inhibitor for human plasma as substrate. In both cases, the K_i-value is 1 μM (fig. 3). This is an inhibitor constant of about the same size as the Km value (0.4 μM) for the reaction between human renin and its protein substrate in

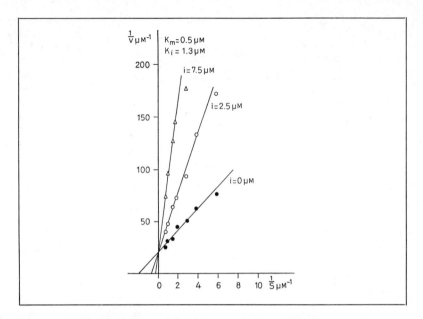

Fig. 3. Competitive inhibition of the reaction between human renin and protein renin-substrate in plasma by Pro-[Phe[5]][Phe[6]]octapeptide. The reaction was performed in the presence of 7.6, 2.5 μM and without (i=o) Pro-[Phe[5]][Phe[6]]octapeptide. Renin concentration was 8×10^{-4} GU/ml and incubation time was 45 min at 37°C and pH 7.5. K_m and K_i values were 0.5 and 1 μM, respectively.

plasma, and the plasma renin activity can be completely blocked. This inhibitor was, therefore, expected to be effective *in vivo*.

However, when a large amount of a saturated solution of the Phe[5]-Phe[6]-octapeptide was injected i.v. to an anaesthetized rat, it was without effect on the blood pressure and was unable to neutralize injected hog renin. Besides a high degradation rate of the peptide *in vivo*, species differences might explain the lack of effect. It is well known that renin exhibits species specificity when reacted with protein renin substrates from other species. In general, renin substrate from primates will only be cleaved by primate renin, and not by renin from other species (table I). However, this species specificity was believed to be due to differences in the substrate molecule in regions other than the cleavage site, since the tetradecapeptide, which is the N-terminal part of the substrate molecule, containing the place of cleavage, is cleaved by all renins (table I).

A study of the effect of the inhibitory peptides on neutralisation of

Table I. Species specificity, K_m (μM) at pH 7.5

Renin	Protein renin-substrate		Tetradecapeptide substrate
	human	rat	
Human	0.5	no	25
Rat	no	3	330
Hog	no	0.4	7

Table II. Inhibitor constant using different renins, K_i values (μM) at pH 7.5

	Human renin		Rat renin, rat substrate	Hog renin, rat substrate
	tetradeca-peptide	human renin substrate		
Pro[pHe^{5}pHe^6]octapeptide	1	1.3	> 75	30
Pro[pHe^6]octapeptide	4	9	>940	>150
Pro$_3$[pHe^6]octapeptide	3	–	>150	>150
Pro[Tyr^6]octapeptide	12	–	>550	>550
Pro[Trp^6]octapeptide	24	–	>330	>330
Pro[pHe^8]octapeptide	71	–	> 85	> 85
Pro$_5$-octapeptide	140	–	>480	>130
Pro$_5$-[D-Leu^6]octapeptide	>500	–	> 1,080	> 1,080

renin from different species revealed that this effect varied greatly (table II). While many inhibitors were active in neutralizing human renin, only Phe^5-Phe^6-octapeptide was effective in neutralizing hog renin (though 30-fold less so); its effect on rat renin was even less, which explains the lack of *in vivo* effect in rats. Studies with monkeys are now under way. The experiments demonstrated, however, that the active sites of renin from primates and non-primates are different, since they have different substrate requirements. This makes it likely that very specific inhibitors for renin can be synthesized, but that they will have different effects in different species.

Finally, a successful use of the inhibitors will be mentioned. As mentioned above, the D-Leu^6-octapeptide was an effective inhibitor at pH 5.5 but was without effect at pH 7.5. This inhibitor was attached to the solid support, sepharose 4-B, by a covalent bond. The support was packed as a column and an impure hog renin sample was applied at pH 5.5. At this pH, renin was bound tightly to the inhibitor, and therefore retained in the column (whereas other proteins were not) when the column was washed with a buffer at pH 5.5. After the impurities had been removed in this way, renin could be liberated again by increasing pH to 7.5, at which the inhibitor was no longer an inhibitor for renin. Hog renin was in this way purified 200-fold from an already purified sample [9], and we now have indications from N-terminal amino acid analysis that the renin is more than 33% pure. One or two contaminating proteins or peptides have still to be removed.

The tetrapeptides and the protease inhibitor pepstatin have also been used for affinity chromatography of renin [9].

It is our hope that further reasearch along these lines will provide us with effective inhibitors which could be used for *in vitro* as well as for *in vivo* studies.

Summary

This review describes some of the characteristics for renin's substrate specificity and also some features of its species specificity. It describes how competitive inhibitors were synthesized and how the solubility was increased in order to make them effective at physiological pH. Their use for *in vitro* and *in vivo* inhibition of renin is discussed and their use for purification of renin demonstrated.

References

1. SKEGGS, L. T.; KAHN, J. R.; LENTZ, K. E., and SHUMWAY, N. P.: The preparation, purification and amino acid sequence of polypeptide renin substrate. J. exp. Med. *106:* 439–444 (1957).
2. KOKUBO, T.; HIWADA, K.; ITO, T.; MEDA, E.; YAMAMURA, Y.; MIZOGUDII, T., and SHIGEZANE, K.: Peptide inhibitors of renin angiotensinogen reaction system. Biochem. Pharmacol. *22:* 3217–3223 (1973).
3. POULSEN, K.; BURTON, J., and HABER, E.: Competitive inhibitors of renin. Biochemistry *12:* 3877–3882 (1973).
4. DIXON, M. and WEBB, E. C.: Enzyme (Longmans, London 1964).
5. WEBB, E. C.: in GRANT and KLYNE Steric aspects of the chemistry and bichchemistry of natural products. Biochem. Soc. Symp. No. 19, p. 90 (Cambridge University Press, Cambridge 1960).
6. SHECHTER, I. and BERGER, A.: On the size of the active site in proteases. I. Papain. Biochem. biophys. Res. Commun. *27:* 157 (1967).

7 PARIKH, I. and CUATRECASAS, P.: Substrate analog competitive inhibitors of human renin. Biochem. biophys. Res. Commun. *54:* 1356–1361 (1973).
8 BURTON, J.; POULSEN, K., and HABER, E.: Competitive inhibitors of renin. Inhibitors effective at physiological pH. Biochemistry *14:* 3892–3898 (1975).
9 POULSEN, K.; BURTON, J., and HABER, E.: Purification of hog renin by affinity chromatography using the synthetic competitive inhibitor [D-Leu^6] octapeptide. Biochim. biophys. Acta *400:* 258–262 (1975).

Dr. K. POULSEN, University Institute for Experimental Medicine, *Copenhagen* (Denmark)

Discussion[1]

GANTEN: Dr. REID, first, if you inject substrate into the brain or into the cerebral ventricles it produces the angiotensin response. So it seems that, at the pH which is present in the brain, the enzyme does react, leading to angiotensin formation, which can be blocked by P113.

Second, how do you reconcile your findings with those reported by Dr. MALVIN at the Houston meeting that P113 does block thirst drive when injected intraventricularly?

REID: The problem with the substrate injections is that the tetradecapeptide has been used. This is not the naturally occurring substrate and until the natural substrate is tested, the question remains open. MALVIN reported that saralasin caused a reduction in the drinking response to water deprivation. It was a very small reduction and, as I remember, barely statistically significant.

GORDON: Dr. REID, you showed us effects on drinking in water deprived animals in which you injected renin only a few minutes before the inhibitor. Did you do any long-term studies with infusion or did you measure vasopressin levels? The time course seems very brief.

REID: I agree that we should look at the effects of saralasin over a longer period of time. However, I am confident that the angiotensin receptors are blocked in a matter of minutes, because the responses to injections of angiotensin are abolished within 2 min of the saralasin injection.

DENTON: I would like to confirm what Dr. REID has said in relation to the drinking situation. We find that the injection of P113 into the third ventricle, at rates of infusion in the conscious animal which completely block the dipsogenic effect of intracarotid angiotensin II, is without influence on drinking caused by water deprivation, by intracarotid 4-molar sodium chloride, or by the thirst which follows rapid eating in the sheep (which could be a volume-induced drinking). A large number of people who are working in this behaviour field have not considered what is

1 Discussion of the papers of REID and POULSEN *et al.*

Discussion

physiologically feasible in relation to what they inject. When they put a nanogram or 100 pg into the brain and produce drinking, that is contained in a microlitre, giving a concentration of anything from 10,000 to 100,000 ng/100 ml. In other words, these neurones get a 'bomb' let off in their environment which is completely outside any physiological range in relation to angiotensin from renin of renal origin. My question, Dr. REID, is in relation to your measurements of angiotensin in the cerebrospinal fluid, which work out to levels of 100 ng/100 ml from your 0.1 Goldblatt units: Have you gone down to 0.01 Goldblatt units, which is getting down to the sort of levels of angiotensin that we find in the cerebrospinal fluid of sodium depleted animals (the highest levels we ever get are 5–10 ng/100 ml)? Do you get drinking then?

REID: No, we have not done it, but I suspect that if we did we would not get drinking.

McDONALD: I would just like to extend what Dr. REID has said about intraventricular renin and angiotensin. Systemically, angiotensin II has the opposite effect to what one would predict if it were a vasopressin-releasing hormone; that is, when given systemically angiotensin II, in a pressor dose causing about 25 mm Hg rise in arterial pressure, produces a diuresis that can be blocked by denervation of arterial baroreceptors. The diuresis is absent also if you do an acute hypophysectomy and give the animal a constant infusion of vasopressin as well as a pressor infusion of angiotensin II. If the systemic renin angiotensin system is really an effective pressor system, it probably works on water by baroreceptor mechanisms to suppress the vasopressin release. It is a 'diuretic' system if one wants to look at it in that way.

PETTINGER: Dr. REID, do you have any studies on the effects of lowering blood pressure. The possibility is that this might have a central effect which could be blocked by P113 or some change in ADH.

REID: No.

VALLOTTON: In relation to what Dr. POULSEN said about the difference of the renin substrates in different species, as evidenced by the effects of the various analogues he reported, I would like to remind you that some years ago with Dr. FAVRE and ROUSSEL-DERUYCK [Biochim. biophys. Acta 302: 102–109, 1973], we showed that the pH optimum of renin activity could depend not so much on the enzyme itself but also on the substrate utilised. Substrates coming from different species may have different pH optima because they have a different tendency to be denatured at a lower pH. We could demonstrate that the synthetic tetradecapeptide was not denatured at a low pH.

POULSEN: The species specificity of these peptides were demonstrated over the whole pH optimum curve for these enzymes, and the values that I gave here refer to the pH optimum for each particular enzyme with the substrate.

MULROW: First, I would like to congratulate Dr. POULSEN on a beautiful demonstration of his work and I think it illustrates that when you start looking at some basic principle, the benefits, not visible when you start, appear later. Already you have been able to use this peptide to purify renin; you have learned something more about substrate specificity, and also about where the specificity may lie in the molecule.

My question is whether you can give reasonable amounts of the peptide, or whether because of all the other proteins in plasma, including substrate, the inhibitor may not work *in vivo*.

POULSEN: Thank you very much for your comment. The question of whether or not inhibitors of this kind will be effective *in vivo* is determined by two things, first of all, the affinity – the K_i value – and, secondly, the amount which we can inject. This is determined by the solubility. In respect to the affinity, I think that the affinities of our peptides are of the same order of magnitude as that of the substrate. The inhibitors from KOKUBU's work are three or four orders of magnitude lower than the inhibitors just presented. This again means that if you have an inhibitor which has the same K_i value as the K_m value of the substrate reaction, an inhibitor concentration equal to the substrate concentration will inhibit about 50% of the renin present. Then comes the question of the solubility. From our calculations, we expect to obtain a concentration of inhibitor peptide about tenfold higher than the substrate concentration, which should be enough. On the other hand, we are just at the beginning of this study and it seems very likely that we will be able to increase the solubility much further.

In relation to the experiments in monkeys, injection of the inhibitors really does inhibit the pressor effect of the injection of human renin, so it is possible to obtain a sufficiently high concentration.

EDWARDS: We have some evidence that reducing the circulating levels of substrate by doing portacaval anastomosis has a blood pressure lowering effect, and that this is linearly correlated – the more you lower the circulating plasma renin substrate level by the portacaval shunt operation, the lower the blood pressure falls. These studies suggest that reducing the amount of substrate available either by reducing hepatic production and release as we have done, or by blocking conversion to angiotensin as Dr. POULSEN is attempting, may well provide a useful approach to the control of blood pressure.

COGHLAN: Dr. REID, I take it that you are fairly sure that the renin you inject does not contain converting enzyme and substrate, and that you have thereby established that these are available. Considering also the experiments that Dr. GANTEN quotes, it seems whether or not it is renin or tetradecapeptide, you put it in and it works, thus establishing that there is both enough renin available, and enough substrate.

Why is it that these dogs do not drink all the time?

REID: We have used two renin preparations: one is the Nutritional Biochemicals hog renin product and the other is one of the Haas dog renin preparations. I do not think that these contain substrate or converting enzyme.

Why do not the dogs drink all the time? I think that the explanation is that there is no brain renin activity at physiological pH.

Section III: Clinical Studies

In STOKES and EDWARDS: Drugs Affecting the Renin-Angiotensin-Aldosterone System. Use of Angiotensin Inhibitors
Prog. biochem. Pharmacol., vol. 12, pp. 145–162 (Karger, Basel 1976)

Angiotensin II Blockade in Normal Man and Patients with Essential Hypertension

Blood Pressure Effects Depending on Renin and Sodium Balance

H. R. BRUNNER, H. GAVRAS, A. B. RIBEIRO and L. POSTERNAK

Departments of Medicine, Centre Hospitalier Universitaire, Lausanne; Boston University School of Medicine, Boston, Mass., and Cornell University Medical College, New York, N.Y.

Contents

Introduction	145
Methods	146
Normal Volunteers	146
Patients with Essential Hypertension	147
Laboratory Determinations	149
Results	149
Normotensive Volunteers	149
Blood Pressure	151
Sodium Balance	151
Plasma Renin Activity	151
Patients with Essential Hypertension	152
Blood Pressure	154
Sodium Balance	157
Plasma Renin Activity	157
Discussion	157
Summary	160
References	161

Introduction

During recent years, the renin-angiotensin system has been more and more implicated in the pathogenesis of clinical hypertensive diseases and has been thought to participate in normal blood pressure regulation. Systematic exploration of this hormonal system in experimental and clinical hypertension has lead to new concepts concerning the mechanisms which sustain normal and elevated blood pressure. Moreover, based on measurements of plasma renin activity, different subgroups of essential hypertension have emerged, each apparently with different physiopathological and clinical characteristics [1]. However, the measurement alone of the different components of this hormonal system cannot suffice to determine whether and to what extent the system participates in blood pressure control, since equal renin levels can exert variable pressor activity depending on their interaction with the sodium volume factor [2]. Specific pharmacological blockade of the angiotensin system may provide a better approach to the problem because it may make it possible to evaluate directly the pressor effect of the system. Such angiotensin II blockade has been used previously to study, both in animals [3–8] and in man [9–14], renovascular, renal and malignant hypertension under various conditions of sodium balance in order to identify the mechanisms sustaining high blood pressure.

In the present study, the angiotensin II inhibitor Sar^1-Ala^8-angiotensin II (saralasin) [15] was administered to normotensive volunteers and to patients with essential hypertension in an effort to extend our previous studies to this large and heterogenous hypertensive population. Blood pressure response to angiotensin II blockade was recorded during sodium depletion and sodium repletion with the patients in supine as well as in upright positions.

Methods

Normal Volunteers

Nine normotensive male volunteers, 22–24 years of age, were included in the study. They were initially placed on a constant diet containing 10 mEq of sodium and 60 mEq of potassium per day for a period of 6 days. Subjects 1, 2, 7, 8 and 9 received chlorthalidone 100 mg p.o. on day 1 and subjects 1, 2 and 8 another 100 mg on day 2. On day 6, after stabilization of the blood pressure in the supine position for at least 60 min, saralasin was infused intravenously at the rate of 10 µg/kg body weight/min for a period of 60–90 min with the subject in the supine position and another 30 min, or until fainting occurred, in the upright position. Blood pressures were monitored at 2- to 5-min intervals using a mercury sphygmomanometer or an automatic blood pressure recorder

(Arteriosonde 1216/1217, Roche). Subsequently, all volunteers were subjected to repletion of sodium for 6 days with a constant diet containing 100 mEq of sodium and 60 mEq of potassium per day. Then on the 6th day of sodium repletion, the infusion of saralasin was repeated under identical conditions.

Blood samples for the determination of plasma renin activity were collected on the 5th day of each period of constant diet with the subject in the upright position. On the day of the saralasin infusion, blood samples for renin measurement were obtained prior to the infusion, 60–90 min following the start of the infusion with the subject still supine and finally after the subject had assumed the erect position for 30 min or until the moment of fainting. 24-hour urine was collected daily throughout the study for the measurement of sodium, potassium and creatinine.

Patients with Essential Hypertension

13 hypertensive patients, 27–64 years of age, were included in the study. All exhibited a diastolic blood pressure of at least 110 mm Hg. All had grade II retinopathy and serum creatinine levels of less than 1.3 mg%. Antihypertensive therapy had been discontinued for at least 2 weeks prior to the study. Basic diagnostic work-up to exclude any primary cause of the hypertension comprised the measurement of plasma and urine electrolytes, blood urea nitrogen concentration, creatinine clearance, screening for phaeochromocytoma and rapid sequence intravenous pyelogram. Furthermore, all except one patient underwent renal arteriography.

The protocol used for the study of the hypertensive patients was identical to the one utilized for the study of the normal volunteers except for the following points: Unlike the normal subjects, every hypertensive patient received a total dose of 200 mg of chlorthalidone during the first two days of the sodium depletion period. In addition, only 9 of the patients were studied after sodium repletion. In 1 patient, blood pressure normalized during sodium repletion and she therefore was not included in further studies. Three other patients exhibited a marked rise in their blood pressure during the first infusion of saralasin, which necessitated a reduction of the dose to 1 μg/kg body weight/min. Two of these patients were subsequently exposed to an additional prolonged period of vigorous sodium depletion. This was achieved by continuing the low sodium diet (10 mEq Na/day) combined with administration of either spironolactone 200–300 mg/day for 16 days (patient D. R.) or a combination of chlor-

Table I. Summary of results in normal volunteers

Subjects	Chlorthalidone (total dose, days 1+2), mg	Mean blood pressure, mm Hg				Na repletion				Cumulative Na balance, mEq	
		Na depletion				control:		angiotensin inhibition		depletion	repletion
		control:		angiotensin inhibition		supine		supine	upright		
		supine		supine	upright						
Responders											
1	200	80		73	(52)	73		77	93	−418	+489
2	200	87		70	(37)	73		83	83	−437	+474
3	0	92		95	(57)	89		89	95	−213	+150
7	100	79		73	(20)	78		84	90	−321	+410
8	200	87		(40)	–	79		81	78	−376	+397
Non-responders											
4	0	75		78	93	75		83	93	− 98	+185
5	0	78		81	84	70		73	80	− 66	+151
6	0	91		93	92	87		90	87	− 12	+130
9	100	100		95	84	87		92	95	−152	+256

Lowest blood pressures measured are given in parentheses. Pressure dropped even further, but could not be determined with accuracy because of subject fainting.
See text for definition of responders and non-responders.

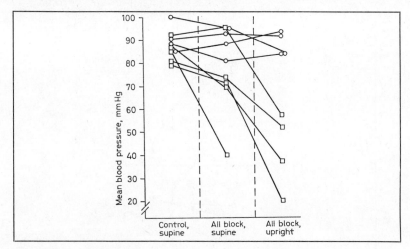

Fig. 1. Nine normotensive volunteers: individual blood pressure responses to saralasin during sodium depletion. 'Responders' (□) exhibit a reduction of mean blood pressure ≧20 mm Hg. O = 'Non-responders' (see text).

thalidone 100 mg/day and spironolactone 200 mg/day for 9 days (patient P. D.). Infusion of saralasin at the rate of 1 µg/kg/min was repeated at the end of this period.

Laboratory Determinations

Plasma renin activity was determined by radioimmunoassay, measuring generated angiotensin I after 3 h of incubation, and the results were expressed as ng/ml/h [16]. Urinary electrolytes were measured by flame photometry. Cumulative sodium balances were calculated for the period of sodium depletion and sodium repletion, respectively. This was done for the normal volunteers by subtracting urinary sodium excretion from sodium intake and correcting the value for insensible losses (5 mEq/day during sodium depletion and 10 mEq/day during sodium repletion). For the hypertensive patients, cumulative sodium balances were not corrected for insensible losses.

Results

Normotensive Volunteers

Table I gives an overview of some findings made in the normotensive volunteers.

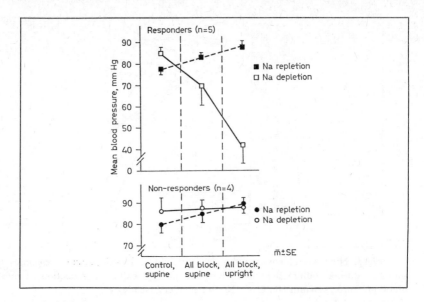

Fig. 2. Normotensive volunteers: mean blood pressure responses to saralasin of 'responders' and 'non-responders' during sodium depletion and sodium repletion.

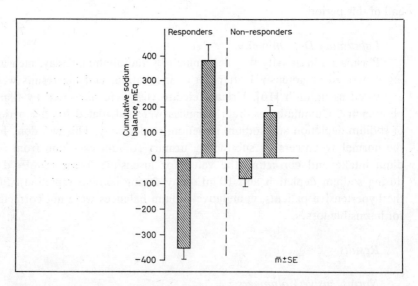

Fig. 3. Normotensive volunteers: cumulative sodium balances calculated for periods of sodium depletion and sodium repletion. It is evident that 'responders' were more vigorously sodium-depleted during the initial six days.

Blood Pressure

Figure 1 depicts the individual mean blood pressure responses (diastolic blood pressure + one third of the pulse pressure) of these 9 subjects to saralasin infusion on the 6th day of sodium depletion. Only one subject exhibited a significant blood pressure reduction after 90 min of angiotensin II inhibition while still supine. In the other 8 subjects, either no change or only a slight trend towards blood pressure reduction occurred during this period. However, when standing up, 4 additional volunteers showed a significant blood pressure fall. In all 5 subjects who developed hypotension, there was clinical evidence of catecholamine oversecretion, such as profuse perspiration, tachycardia and nausea. The 5 together are arbitrarily labelled 'responders' (depicted as open squares). The 4 remaining patients (depicted as open circles) showed no change in blood pressure even when standing up. Accordingly, they are called 'non-responders'.

In figure 2, the mean blood pressure responses of the 2 groups during 2 consecutive saralasin studies can be seen. During sodium repletion, neither the responders nor the non-responders exhibited any significant change in blood pressure.

Sodium Balance

In figure 3, the cumulative sodium balances for the 2 consecutive study periods and the 2 groups are shown. The 'responders' had been significantly more sodium-depleted prior to the first saralasin study than the 'non-responders'. Negative cumulative sodium balance during the first 6 days was -353 ± 40.3 mEq for the responders as compared to -82 ± 29.3 mEq for the non-responders. From table I, it is evident that this difference is due to a more intensive diuretic treatment of the subjects making up the responder group.

Plasma Renin Activity

In figure 4, the results of the renin determinations are summarized for the two groups. On day 5, upright plasma renin activity of the responders reached 41.4 ± 9.4 ng/ml/h. In the supine position on the morning of day 6, plasma renin activity was 32.7 ± 7.2 ng/ml/h. Subsequent angiotensin II blockade resulted in a steep increase to 133.3 ± 42.9 ng/ml/h and assuming the erect position did not induce any further significant change since renin levelled off at 116.3 ± 50.7 ng/ml/h. During the same period, the non-responders showed essentially the same pattern of response, though at a lower level. Thus, plasma renin activity changed from

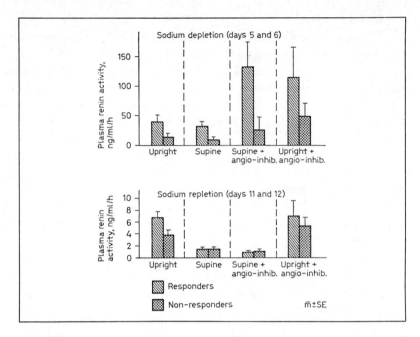

Fig. 4. Normotensive volunteers: response of plasma renin activity to angiotensin II blockade. Saralasin stimulates renin secretion only if baseline secretion is high.

14.7 ± 5.1 ng/ml/h the 5th day to 9.3 ± 4.6 ng/ml/h on the morning of the 6th day with the subject in the supine position. Saralasin infusion stimulated renin to 26.8 ± 21.1 ng/ml/h, and assuming the erect position resulted in a further increase to 50 ± 21.7 ng/ml/h. In the sodium-replete state, both groups showed similar renin characteristics. In contrast to the response observed during sodium depletion, in the sodium-replete state, when renin values were rather low, infusion of saralasin did not seem to have a stimulatory effect on renin release.

Patients with Essential Hypertension

Table II reviews some of the characteristics of the hypertensive patients included in the study.

Table II. Summary of results in hypertensive subjects

Patients	Renin classification according to nomogram	Mean blood pressures, mm Hg							Cumulative Na balance, mEq	
		Na depletion				Na repletion			depletion	repletion
		control: supine	angiotensin inhibition supine	angiotensin inhibition upright		control: supine	angiotensin inhibition supine	angiotensin inhibition upright		
Group I										
P.T.	NL	137	102	87		130	150	170	−112	+254
H.B.	Hi	112	78	<70		110	115	95	−397	+533
A.F.	Hi	127	83	<70		112	108		−290	+259
M.F.	Hi	115	97	88		113	117	133	−180	+199
S.L.	Hi	103	83	<70		90			−217	
Group II										
H.Br.	NL	115	122	102		112	130	143	−344	+402
M.D.	Lo	120	120	135		130	170		−168	+322
N.T.	NL	122	114	138		118	143	152	−143	+316
K.T.	Lo	100	104			117	121	143	−172	+333
R.W.	NL	105	110	123		107	122	143	− 87	+444
Group III		Initial depletion				Prolonged Na depletion				Further depletion
M.G.	Lo	134	157			120	103	88	−103	−244
D.R.	Lo	113	142	162		102	100	95	−223	−520
P.D.	Lo	117	137						− 89	

NL = Normal renin; Hi = high renin; Lo = low renin.

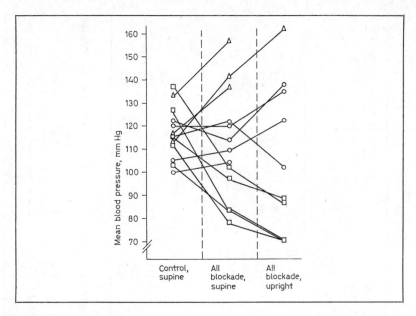

Fig. 5. Essential hypertension: individual blood pressure responses of 13 patients to saralasin during sodium depletion: group I (□); group II (○); group III (△) (see text).

Blood Pressure

The individual blood pressure responses of all 13 patients to saralasin infusion on the 6th day of sodium depletion are depicted in figure 5. Three response patterns emerged. Five patients exhibited a marked and significant blood pressure reduction during angiotensin II inhibition; they are depicted by open squares and constitute group I. Five other patients, depicted by open circles, did not show any significant change in blood pressure during angiotensin II blockade, though a slight but not significant increase in blood pressure could be seen in some of the patients when assuming the erect position. These comprise group II. The remaining 3 patients, depicted by open triangles, showed a marked agonistic blood pressure effect of saralasin, resulting in significant blood pressure elevation which necessitated a reduction of saralasin to 1 μg/kg/min and even discontinuation of the infusion. These latter patients make up group III.

Figure 6 shows the pattterns of the blood pressure responses of the 3 different groups. While blood pressure of group I fell during sodium de-

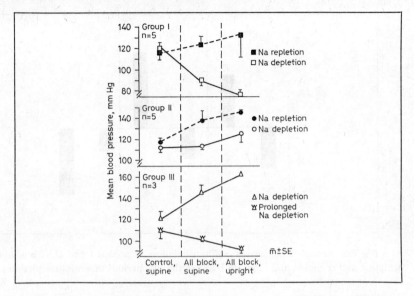

Fig. 6. Essential hypertension: mean blood pressure responses to saralasin of group I, II, and III patients during sodium depletion and sodium repletion or prolonged sodium depletion (group III). Note particularly change of response in group III following prolonged sodium depletion.

pletion from 119 ± 5.9 to 90 ± 4.5 mm Hg in response to angiotensin II blockade, blood pressure reduction could not be induced when the same patients were sodium repleted. Group II patients exhibited no change in blood pressure during the first infusion of saralasin (blood pressure changed from 112 ± 4 to 114 ± 3 mm Hg while lying). However, when the infusion was repeated after sodium repletion, blood pressure rose significantly from 117 ± 4 to 137 ± 9 mm Hg ($p<0.05$) with the patients lying, and it increased further to 145 ± 2, when they stood up. Finally, in group III, the infusion of the blocker induced a significant rise in blood pressure from 121 ± 6 to 145 ± 6 mm Hg ($p<0.01$) despite our standard sodium depleting regimen. This elevation persisted even when the infusion rate was lowered to 1 μg/kg/min. The one patient who was then asked to stand up showed a further elevation of mean blood pressure to 162 mm Hg. In the light of the results obtained in group II, it was considered unwise to study the patients of group III in the sodium-replete state. Therefore, following the first infusion, sodium depletion was prolonged

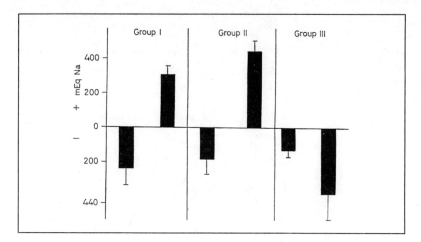

Fig. 7. Essential hypertension: sodium balances in groups I and II after sodium depletion and repletion and in group III after the two periods of sodium depletion.

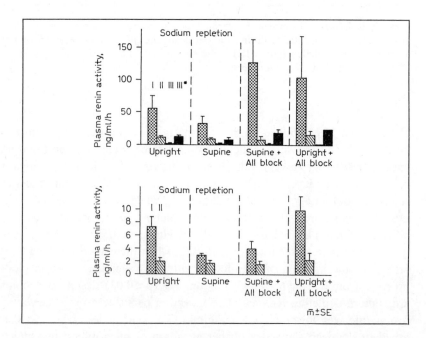

Fig. 8. Essential hypertension: mean responses of plasma renin activity to saralasin administration of the 3 groups during sodium depletion and repletion. I = group I; II = group II; III = group III during initial sodium depletion, and III* = group III during prolonged sodium depletion.

and intensified using diuretic therapy. After this second period of sodium depletion, to which 2 patients were exposed, their mean blood pressure fell from 111 ± 9 to 102 ± 2 mm Hg while supine, and this reduction was further accentuated to 92 ± 4 mm Hg when the patients rose.

Sodium Balance

At the end of the initial salt-depletion period, sodium balance was most negative in group I (–239 ± 48.2 mEq). Group II had lost 182 ± 43.1 mEq and group III only 138.3 ± 42.5 mEq (fig. 7). The two patients of group III, who subsequently underwent more vigorous sodium depletion, lost an additional 382 ± 138 mEq. Cumulative sodium balance during sodium repletion attained + 311 ± 25 mEq for group I and + 444.2 ± 25 mEq for group II.

Plasma Renin Activity

The renin results of the different groups are summarized in figure 8. Patients included in group I exhibited clearly higher renin levels than those of group II and this held as well during sodium depletion as during sodium repletion. Indeed, 4 of the 5 patients included in group I had high renin levels according to our nomogram, relating plasma renin activity to concomitant 24-hour urinary sodium excretion [1]. All patients of group III exhibited abnormally low renin values. Saralasin infusion induced a marked increase in renin levels of group I when sodium-depleted. In the other 2 groups and during sodium repletion, angiotensin II inhibition did not appear to exert any marked stimulating effect on renin secretion.

Discussion

From these studies of normotensive volunteers and patients with essential hypertension, one main concept seems to emerge: the response of the blood pressure to the infusion of the angiotensin II antagonist depends primarily on the concomitant state of sodium balance. Thus, out of nine normal volunteers, only those five who had been most vigorously sodium-depleted exhibited a significant blood pressure reduction during angiotensin II blockade by saralasin. Actually, it appears that a cumulative sodium loss of at least 160–200 mEq was necessary to achieve that result. With one exception, low sodium diet alone was not sufficient to achieve this degree of sodium depletion, so that administration of diuretics was necessary to enhance excretion of sodium. Moreover, following sodium repletion, none of the volunteers responded to angiotensin II blockade with a significant change in blood pressure. Accordingly, angiotensin II seems to participate directly in the maintenance of normal blood pressure only under conditions of considerable sodium depletion.

Catecholamine hypersecretion as evidenced by such signs as pallor, sweating, tachycardia, was incapable of sustaining blood pressure during angiotensin II inhibition and marked sodium depletion, possibly because its vasopressor effect was impaired in the absence of the permissive action of the sodium ion [17]. Thus, during severe hypovolaemia, the renin-angiotensin system appears to represent a key mechanism maintaining normal blood pressure even in the face of a normally functioning catecholamine and autonomic nervous system.

A similar, though more complex pattern of blood pressure responses to angiotensin II inhibition was observed in the patients with essential hypertension. If a critical degree of sodium loss could be achieved, blood pressure fell significantly during infusion with saralasin, whereas with insufficient depletion, or upon repletion of sodium, the infusion of saralasin produced either no antagonistic or even an agonistic effect with variable degree of blood pressure elevation. Group I patients lost a sufficient amount of salt with our standard sodium-depleting regimen to cause a marked oversecretion of renin. Under these conditions, infusion of saralasin produced a marked fall in blood pressure. Group II patients under the same regimen lost less sodium which resulted only in a mild compensatory stimulation of renin secretion; infusion of the inhibitor in these cases produced practically no change in blood pressure. Finally, group III patients lost even less sodium than the other 2 groups under the same regimen. The induced depletion was not sufficient to stimulate renin secretion. In these patients, saralasin, by its inherent slight agonistic effect, which may have been enhanced by a permissive action of the sodium ion on vascular receptor activity, produced a marked increase in blood pressure. This hypertensive effect of the inhibitor was similar to the one observed in group II patients following sodium loading. However, when the group III patients were further sodium-depleted by prolonged and vigorous treatment with diuretics, angiotensin II inhibition induced a clear blood pressure reduction.

These results suggest that if a degree of sodium depletion is achieved which is sufficient to induce marked stimulation of renin secretion, the angiotensin component seems to participate directly in the maintenance of essential hypertension. The main peculiarity of the group III patients was their initial state of apparently excessive sodium content and their relative inability to become sodium-depleted under a standard depleting regimen, whereas their capacity to stimulate renin secretion under appropriate conditions was not abolished. The impression that these patients exhibit an important defect in sodium handling and/or some degree of sodium retention is further supported by the observation that their blood pressure response following initial sodium depletion was similar to the one observed in group II patients after sodium repletion.

The reciprocal relationship between sodium dependency and renin dependency observed in the present investigation is in perfect agreement with results obtained previously in studies of normotensive rats [4] and animals with two or one-kidney renovascular hypertension [4, 5], as well as in studies of human hypertension of the renovascular, advanced or malignant type, utilizing saralasin [10] or an inhibitor of the conversion from angiotensin I to angiotensin II [12]. It is conceivable that most differences among the various types of hypertension could be expressed in terms of their tendency to retain or inability to excrete sodium. It appears that patients who need to eliminate the biggest amount of sodium to normalize their blood pressure are those who actually excrete the smallest amount during a standard depleting regimen, whereas the ones who readily lose sodium with such a regimen appear to have retained less sodium in the first place. This tendency, variable in each individual case, covers the whole spectrum between the two extremes and appears to be the factor determining the degree of renin dependency.

Classification of patients with essential hypertension into high renin, normal renin and low renin groups as a means of predicting response to treatment may turn out to be of relatively more limited practical value than has been suggested [18]. More meaningful would seem the direct determination of the degree of renin dependency under different conditions of sodium balance, which may not necessarily be indicated by the levels of plasma renin activity unless they fall into one of the two extremes of very low or very high. Thus, in the present study, one normal renin patient behaved similarly to the high renin group whereas other normal renin cases were indistinguishable from those with low renin in terms of blood pressure response to blockade of angiotensin II.

Using the approach of pharmacologic angiotensin II blockade by saralasin for diagnostic evaluation of hypertensive patients may provide additional potentially very useful information. Not only the presence or absence of renin dependency can be detected but, at the same time, the presence and even the degree of overactivity of the sodium pressor factor can be identified via the inherent agonistic effect of saralasin. Since the sodium-volume component appears most frequently to play an important role in sustaining high blood pressure levels, this latter information could prove to be of great practical value in choosing more specific therapy for the primary disorder.

The present data demonstrate once more that angiotensin II inhibition can stimulate renin secretion. A relationship appears to exist between the potency of the renin-stimulating effect of angiotensin II blockade and the preexisting level of renin secretion. Thus, if the baseline renin value is high, angiotensin II inhibition induces a marked increase in renin secretion. If, however, renin is low, it may rise only little in response to the infusion of saralasin or, as has been observed in all our sodium replete subjects, renin stimulation may actually be altogether absent despite an equal dose of saralasin. These observations may suggest that negative feedback inhibition of renin secretion by angiotensin II becomes operative only when circulating angiotensin II levels are elevated.

Taken together, it appears that in all patients with essential hypertension, as in other forms of experimental and human hypertension as well as in the normotensive state, sodium balance determines the degree of participation of the renin angiotensin system in sustaining blood pressure. The main difference between hypertensive patients with different renin levels appears to be their variable capacity to excrete or retain sodium, while all are capable of showing enhanced renin release under appropriate stimulation.

Summary

1) Saralasin was administered to 9 normotensive volunteers and 13 patients with essential hypertension after sodium depletion and sodium repletion.

2) In standing normotensive volunteers, angiotensin II inhibition induced significant hypotension if previously a cumulative sodium loss of at least 160–200 mEq had been induced.

3) In patients with essential hypertension, saralasin infusion induced either blood pressure reduction, no change or even significant blood pressure increase, depending on the prevailing state of sodium balance.

4) Following vigorous and prolonged sodium depletion induced by low sodium diet, with chlorthalidone and spironolactone, blood pressure became renin-dependent even in those patients who intially had exhibited a hypertensive response to saralasin, suggesting that under appropriate conditions, renin can play an active pressure role in all patients with essential hypertension.

5) Saralasin administration to patients with essential hypertension may not only be useful for recognizing renin dependency but may also, via the slight intrinsic agonistic effect of the compound, permit identification of overactivity of the sodium factor.

References

1 BRUNNER, H. R.; SEALEY, J. E., and LARAGH, J. H.: Renin subgroups in essential hypertension. Further analysis of their pathophysiological and epidemiological characteristics. Circulation Res. *32–33:* suppl. I, pp. 99–109 (1973).
2 BRUNNER, H. R. and GAVRAS, H.: Clinical implications of renin in the hypertensive patient. J. Am. med. Ass. *233:* 1091 (1975).
3 BRUNNER, H. R.; KIRSHMANN, J. D.; SEALEY, J. E., and LARAGH, J. H.: Hypertension of renal origin. Evidence for two different mechanisms. Science *174:* 1344–1346 (1971).
4 GAVRAS, H.; BRUNNER, H. R.; VAUGHAN, E. D., and LARAGH, J. H.: Angiotensin-sodium interaction in blood pressure maintenance of renal hypertensive and normotensive rats. Science *180:* 1369–1372 (1973).
5 GAVRAS, H.; BRUNNER, H. R.; THURSTON, H., and LARAGH, J. H.: Reciprocation of renin-dependency with sodium volume dependency. Science *188:* 1316–1317 (1975).
6 KRIEGER, E. M.; SALGADO, H. C.; ASSAN, C. J.; GREEN, L. L. J., and FERREIRA, S. H.: Potential screening test for detection of overactivity of renin-angiotensin system. Lancet *i:* 269–271 (1971).
7 MILLER, E. D.; SAMUELS, A. J.; HABER, E., and BARGER, A. C.: Inhibition of angiotensin conversion in experimental renovascular hypertension. Science *177:* 1108–1109 (1972).
8 BUMPUS, F. M.; SEN, S.; SMEBY, R. R.; SWEET, C.; FERRARIO, C. M., and KHOSLA, M. C.: Use of angiotensin antagonists in experimental hypertension. Circulation Res. *32–33:* suppl. I, pp. 150–158 (1973).
9 BRUNNER, H. R.; GAVRAS, H.; LARAGH, J. H., and KEENAN, R.: Angiotensin II blockade in man by sar^1-ala^8-angiotensin II for understanding and treatment of high blood pressure. Lancet *ii:* 1045–1048 (1973).
10 BRUNNER, H. R.; GAVRAS, H.; LARAGH, J. H., and KEENAN, R.: Hypertension in man. Exposure of the renin and sodium components using angiotensin II blockade. Circulation Res. *34–35:* suppl. I, pp. 35–46 (1974).
11 BRUNNER, H. R. and GAVRAS, H.: The role of renin and sodium in high blood

pressure regulation; in BERGLUND, HANSSON and WERKÖ Pathophysiology and management of arterial hypertension, pp. 32–42 (Mölndal, Hässle 1975).
12 GAVRAS, H.; BRUNNER, H. R.; LARAGH, J. H.; SEALEY, J. E.; GAVRAS, I., and VUKOVICH, R. A.: An angiotensin converting enzyme inhibitor to identify and treat vasoconstrictor volume factors in hypertensive patients. New Engl. J. Med. *291:* 817–821 (1974).
13 STREETEN, D. H. P.; ANDERSON, G. H.; FREIBERG, J. M., and DALAKOS, T. G.: Use of an angiotensin antagonist (saralasin) in the recognition of 'angiotensinogenic' hypertension. New Engl. J. Med. *296:* 657–662 (1975).
14 JOHNSON, J. G.; BLACK, W. D.; VUKOVICH, R. A., *et al.*: Treatment of patients with severe hypertension by inhibition of angiotensin-converting enzyme. Clin. Sci. mol. Med. *48:* 53S–56S (1975).
15 PALS, D. T.; MASUCCI, F. D.; SIPOS, F., and DENNING, G. S.: A specific competitive antagonist of the vascular action of angiotensin II. Circulation Res. *29:* 664–672 (1971).
16 SEALEY, J. E.; GERTEN-BANES, J., and LARAGH, J. H.: The renin system variation in man measured by radioimmunoassay or bioassay. Kidney int. *1:* 240–253 (1972).
17 RAAB, W.; HUMPHREYS, R. J.; MAKOUS, N.; DEGRANDPRE, R., and GIGEE, W.: Pressor effects of epinephrine, norepinephrine and desoxycorticosterone acetate (DCA) weakened by sodium withdrawal. Circulation *6:* 373–377 (1952).
18 LARAGH, J. H. (ed.): Hypertension manual (Yorke Medical Books, New York 1973).

H. R. BRUNNER, MD, Department of Medicine, Centre Hospitalier Universitaire, *CH–1011 Lausanne* (Switzerland)

In STOKES and EDWARDS: Drugs Affecting the Renin-Angiotensin-Aldosterone System. Use of Angiotensin Inhibitors
Prog. biochem. Pharmacol., vol. 12, pp. 163–169 (Karger, Basel 1976)

The Role of Renin in the Control of Blood Pressure in Normotensive Man

PATRICK J. MULROW and ROBERT NOTH

Department of Medicine, Yale University School of Medicine, New Haven, Conn.

Contents

Introduction	163
Methods	163
Results	166
Normal Subjects	166
Cirrhotic Patients	167
Discussion	167
Summary	169
References	169

Introduction

The renin-angiotensin system influences blood pressure in certain hypertensive states [1–3]. Its role in the control of blood pressure in normotensive states is not clear. The purpose of the present investigation was to study the dependence of blood pressure on the renin-angiotensin system in healthy, normal subjects in physiological states of normal and high renin, and in patients having cirrhosis with ascites – a known condition of high renin. Infusions of saralasin (P113) were used to block the renin-angiotensin system in order to study its role in the regulation of blood pressure.

Methods

Six human volunteers on either a normal sodium diet containing approximately 170 mEq/day, or a low sodium diet of 10 mEq/day were

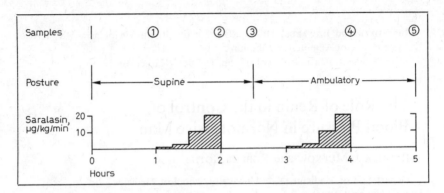

Fig. 1. A diagram showing the experimental design of the P113 infusions into normal subjects. The infusions began at 1 µg/kg/min and were increased at 15-min intervals to 2, 10, and finally 20 µg/kg/min.

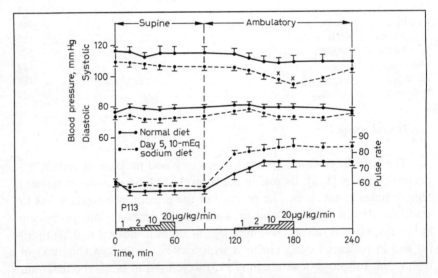

Fig. 2. Blood pressure and pulse rate of six normal subjects before, during and after 60-min infusion of P113 at four different dose rates. Each rate was infused for 15 min. Supine indicates that the subjects were supine for at least 1 h before, during and after infusion. Ambulatory indicates the time period during which the subjects were in erect posture and walking intermittently. The solid line indicates the results from subjects on a normal sodium diet while the interrupted line shows the results in the subjects during the fifth day of a 10-mEq sodium diet. Means and SE are shown. ×Indicates orthostatic symptoms in 3 of 6 subjects.

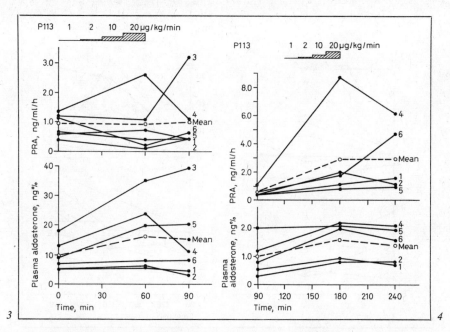

Fig. 3, 4. Showing the individual PRA and plasma aldosterone levels before, at the end of 60 min of P113 infusion and after 30 min post-infusion in the subjects supine (fig. 3) and ambulatory (fig. 4). The interrupted line represents the mean of the results. The numbers indicate the results from an individual patient.

studied in the morning. The subjects were supine for 1 or 2 h before starting the P113 infusion. Blood pressure was recorded every 2–5 min during the control, P113 infusion and the post-infusion or recovery periods. Blood for urea nitrogen, electrolytes, plasma renin activity, aldosterone and cortisol measurements was withdrawn at the end of each period. After the supine period, the subjects became ambulatory. After about 30 min, the P113 infusion was restarted and continued for 1 h. Each time the infusion was stopped, a post-infusion observational period lasted for 1 h. The design of these experiments is shown in figure 1.

Six cirrhotic patients with ascites resistant to bed rest and diuretic therapy were studied only in the supine position. Diuretic therapy was stopped two or more days before the study and the protocol was generally similar to that shown in figure 1 for the supine position. Plasma renin activity (PRA) and plasma aldosterone were measured by radioimmunoassay methods [4].

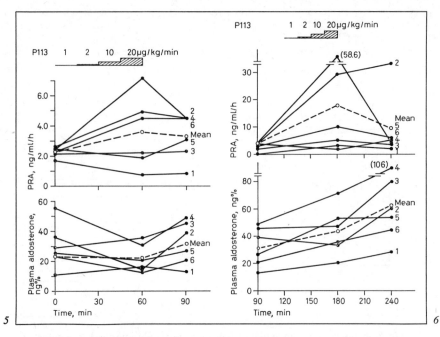

Fig. 5, 6. Showing the individual PRA and plasma aldosterone levels before, at the end of 60 min of P113 infusion and after 30 min post-infusion in the subjects on a low sodium diet supine (fig. 5) and ambulatory (fig. 6). The interrupted line represents the mean of the results. The numbers indicate the results from an individual patient.

Results

Normal Subjects

The effects of P113 on blood pressure and pulse rate in the normal subjects while they were on a normal or low sodium diet are shown in figure 2. P113 infusion occurred during the fifth day of the 10-mEq sodium diet. In subjects on a normal sodium diet, P113 had no effect on blood pressure in either the supine or standing position. Pulse rate rose after subjects became ambulatory and did not increase further during the P113 infusion.

In the subjects on the 10-mEq sodium diet, P113 had no effect upon supine blood pressure, but did cause a significant drop in systolic blood pressure during ambulation. Three of the six subjects had orthostatic symptoms. Of note is the failure of the pulse rate to rise further despite the fall in blood pressure.

With the subjects on a normal diet in the supine position, PRA showed no significant change, while plasma aldosterone rose in three and remained unchanged in three (fig. 3). In these same subjects during the ambulatory period, PRA tended to rise slightly as did aldosterone levels (fig. 4). In the subjects on a low sodium diet in the supine position, PRA rose in three of the subjects, remained constant in two, and fell in one, and aldosterone levels tended to drop (fig. 5). In these same subjects who were ambulatory, the infusion of P113 did not block the rise in aldosterone that occurs with ambulation (fig. 6).

Cirrhotic Patients

In only two of the six cirrhotic patients with ascites resistant to diuretic therapy was there a significant drop in systolic blood pressure during the infusion of P113 (table I). Both of these patients had elevated baseline renins, but other patients who did not respond also had elevated PRA (table II); thus, an elevated baseline renin did not necessarily imply that the subject would respond to an infusion of P113. Plasma aldosterone level showed no consistent changes during the P113 infusion (table III). There was no increase in sodium excretion during the P113 infusion; in fact, there was a slight fall.

Discussion

In a normal supine man on a normal sodium intake, P113 had no effect on blood pressure or PRA and an inconsistent effect on plasma aldosterone level. In normal supine man on a low sodium diet, P113 had no effect on blood pressure and plasma aldosterone levels. Only after ambulation on a low sodium diet were we able to detect a physiological role for angiotensin II in blood pressure control. Surprisingly, P113 did not lower plasma aldosterone levels.

In only two cirrhotic patients with ascites did P113 cause a significant drop in blood pressure during the supine position, suggesting that the renin-angiotensin system was playing a role in controlling the blood pressure in these patients. The absolute level of PRA just before the infusion did not predict who would respond, but the two that did respond had high baseline levels. Again, there was no consistent effect upon aldosterone

Table I. Systolic blood pressure (BP, mm Hg) in the control (Pre) and during the P113 infusion (During) and in the post-infusion period (Post). The BP are the mean of the multiple BP

Patient No.	Pre	During	Post
1	106.8	106.6	105.8
2	100.1	85.0	102.0
3	102.8	105.2	107.6
4	124.0	123.0	124.5
5	126.5	138.4	138.6
6	109.1	99.6	104.5

Table II. Plasma renin activity (ng/ml/min) in periods described in table I. Plasma renin activity was measured at the end of each period

Patient No.	Pre	During	Post
1	4.11	4.31	3.89
2	3.14	7.39	3.92
3	1.46	0.68	0.54
4	4.53	34.8	18.7
5	0.62	1.57	1.10
6	3.89	6.18	5.92

Table III. Plasma aldosterone measurements (ng/dl) were analyzed in an aliquot of the same blood specimen collected for measurement of plasma renin activity

Patient No.	Pre	During	Post
1	21.5	37.3	48.9
2	66.6	66.7	55.6
3	20.7	33.4	24.2
4	116.9	81.9	121.6
5	11.5	19.3	13.9
6	50.3	34.0	33.7

levels. These data suggest that either P113 had a mild agonistic effect upon aldosterone production, or that other factors regulated aldosterone plasma levels during the experiments. Furthermore, our results suggest that the role of the renin-angiotensin system in regulating blood pressure in normotensive states is relatively small, or else its absence can be compensated by other blood pressure regulating mechanisms.

Summary

In order to study the role of renin in regulating blood pressure in normotensive states, saralasin (P113) was infused into normal subjects and patients with cirrhosis of the liver and ascites. In normal subjects on a normal sodium intake, P113 infusion had no effect on blood pressure. Only after the combined stress of a low sodium diet and the upright position did P113 lower the blood pressure. In two of the six cirrhotic patients, P113 caused a significant decrease in BP in the supine position. There was no consistent effect of the P113 infusion on plasma aldosterone or plasma renin activity in the normal or cirrhotic subjects.

References

1 PALS, D. T. and MASUCCI, F. O.: Plasma renin and the antihypertensive effect of 1-Sar-8-Ala-angiotensin II. Eur. J. Pharmacol. 23: 115–119 (1973).
2 BUMPUS, F. M.; SEN, S.; SMEBY, R., et al.: Use of angiotensin II antagonists in experimental hypertension. Circulation Res. 32: suppl. I, pp. 1–150 (1973).
3 PEACH, M. J. and CHIU, A. T.: Stimulation and inhibition of aldosterone biosynthesis in vitro by angiotensin II and analogs. Circulation Res. 35: suppl. I, pp. 1–7 (1974).
4 FORMAN, B. H.; FERNANDEZ-CRUZ, A., jr.; KUCZALA, Z. J., and MULROW, P. J.: Effects of an angiotensin inhibitor on adrenocortical response to angiotensin II, ACTH, and K. Am. J. Physiol. 229: 1713–1717 (1975).

Dr. P. J. MULROW, Chairman, Department of Medicine, Medical College of Ohio, *Toledo, Ohio* (USA)

Discussion[1]

MAXWELL: Dr. MULROW, in regard to the experimental design, you said at the beginning that the subjects were 'ambulatory' for several hours. During that period of ambulation they evidently had an incremental infusion of saralasin taking an hour. Now, if they were standing quietly at that time, could not the fall in blood pressure have simply been what one sees in normal individuals with passive standing for an hour? In other words, could that have affected the results? I am trying to find out, technically, how they kept walking around and got the infusion and blood pressure measurements all at the same time.

MULROW: They had an infusion in the room, and they could walk around the room. Meanwhile, the blood pressure was being taken and the infusion was going in one arm. Clearly, the subjects on the normal sodium intake did not have a drop in blood pressure in the supine position, and other subjects we have looked at on a low-sodium diet, without being infused with saralasin, did not have this kind of drop in blood pressure.

FERRARIO: Dr. BRUNNER, did you measure volumes in the group of responders and non-responding patients?

BRUNNER: No.

ULRYCH: Do you have any data on the fluid balance, such as blood volume or extracellular fluid volume? Referring to an article called *Volume Dependent Essential and Steroid Hypertension* by DUSTAN et al. [Am. J. Cardiol. *31:* 606–615, 1973], careful analysis of their data indicates that, during sodium depletion, the blood pressure is dependent on the product of initial blood volume and plasma renin – the more the product of plasma renin and blood volume, the less the blood pressure drop these subjects had upon volume depletion by a low salt diet.

BRUNNER: We do not have volume measurements. However, you may have noticed that with dietary sodium depletion, we have practically no blood pressure response. Only in the group 3 patients did we observe a slight pressure decrease, which was not even statistically significant.

[1] Discussion of the papers of BRUNNER et al. and MULROW and NOTH.

ULRYCH: I would say that they may have low blood volume, and they may become sensitive because of that.

BOYD: Dr. MULROW, you showed very nicely that there is often an increase in plasma aldosterone during sodium deficiency while you infuse the saralasin, but at the same time, in many of those experiments, there is quite a profound fall in blood pressure, and therefore, I would presume, a profound fall in aldosterone clearance rate.

Have you been able to extrapolate this change in *plasma* aldosterone to secretion rate? Or have you done any direct studies on aldosterone secretion rate?

MULROW: No, we have not.

BRAVO: Dr. MULROW, with regard to studies in the normal individuals on a low sodium diet, was there a positive potassium balance?

MULROW: We did not do accurate balance studies to answer that – we only checked the urines in the last couple of days to see if sodium excretion got down to intake.

BRAVO: Well, do you think that one of the reasons for the inability to block the effect of circulating angiotensin II is potassium retention?

MULROW: While I can not answer that point, I can say that the plasma levels of potassium did not change significantly during the infusion of P113, so it was not that potassium went up and offset any effect of P113; nor was there any change in plasma cortisol levels, so the effect probably was not ACTH-dependent.

GROSS: I wonder whether anybody has ever given these patients on low-sodium diet α-adrenergic blockers to determine what is really contributed to the maintenance of blood pressure by angiotensin and what is contributed by other vasoconstrictors?

RAFTOS: Would you also care to speculate as to why they do not demonstrate a tachycardia?

MULROW: I think that is a very important question, Dr. GROSS. What you are saying is that it is well known clinically that subjects depleted of sodium, by diuretics and otherwise, are very sensitive to adrenergic inhibitors or blockers – no question about that. We see the same phenomena using P113, and I think it is very important from the standpoint of using these angiotensin II inhibitors in the diagnosis of renovascular hypertension.

I think the studies that Dr. BRUNNER presented today showed that you can get the same kind of dependence even in essential hypertension. Now the question is, if you go along the dose-response curve, can you show more dependence on the renin system in the patient with renovascular hypertension? Why do they not get the tachycardia? We do not have any data, but one possibility would be by blocking some of the effects of angiotensin II on the adrenal gland, or of adrenergic nerves on the adrenal gland, thus preventing the release of catecholamines.

BRUNNER: I think that tachycardia is absent with a small fall in blood pressure, but we have regularly seen in our normal subjects, who were very much sodium-depleted, that when their blood pressure dropped, they got tachycardia. However, their diastolic pressures were as low as 30 or 40 mm Hg.

McDONALD: I would like to ask Dr. BRUNNER whether or not the patients who became responders after salt depletion, either with the chlorthalidone or sodium restriction, became normotensive.

BRUNNER: No, with sodium depletion alone they were still hypertensive. It is true that many low-renin hypertensives will lower their blood pressure in response to sodium depletion, but those included in the study did not.

STREETEN: I was very interested in Dr. MULROW's aldosterone results in normal subjects during sodium depletion and ambulation. It struck me that the slope of rise of plasma aldosterone was really quite gentle, in comparison with what happens to our normal subjects on a low-sodium diet, when they stand for 2 h; very frequently their plasma aldosterone might rise from 15 to 75 or even 90 ng%. Do you have any control data on the same subjects to show whether, in fact, the saralasin may have been blocking an increase of aldosterone in response to the upright posture, which was obscured by the way in which you showed the results?

MULROW: No, Dr. STREETEN. That is certainly a good question, and one that has concerned us. It is possible that the slope did change. But the thing that we cannot lose sight of though, is this: if there really was a dependence, a total dependence – as we have been thinking over the years – you would have expected it to drop even more rather than showing a gentle rise.

ROBERTSON: On that last phrase of Dr. MULROW – maybe he was being extravagant, or maybe I misheard him – but he said that he had been supposing for many years that there was total dependence of aldosterone on angiotensin. Did you really mean to say that?

MULROW: No, I really did not mean to say it that way – but rather, that on a low-sodium diet, provided that all other factors were kept constant, it has been thought by some people that renin was the main mechanism by which sodium depletion increased aldosterone secretion.

PETTINGER: We were interested in the same thing that Dr. GROSS alluded to a moment ago, and that is the possibility that we could characterize α-adrenergic constrictor versus angiotensin vasoconstrictor mechanisms in a quantitative way by the use of selective antagonists. We started with animal studies but ran into a rather complicated ball-game. The background for the problem was summarized by Dr. LANGER (Proc. Soc. of Hypertension, Sydney 1976). Far more blood pressure lowering occurred when we blocked the α-adrenergic receptors with phentolamine or dibenzyline than what one would have anticipated. I think we must have been blocking the α-adrenergic inhibitory receptor mechanism on the terminal sympathetic neuron, thus increasing the rate of release of norepinephrine. This mixed catecholamine would in turn activate a vasodilatory β-receptor mechanism, which is then unopposed by the blocked α-receptor postsynaptically. I suggest a word of caution about jumping into the use of α-blocking drugs in the same way as we are using angiotensin antagonists. The picture appears to be far more complicated.

STOKES: It is quite clear from these two studies that before we start to move on to questions of the hypertensive state, we must work out what are the adequate conditions for demonstrating an antagonist effect on blood pressure in normal volunteers – and I would congratulate Dr. BRUNNER on his beautiful studies, designed specifically to reveal those conditions in regard to sodium balance.

Dr. MULROW, I do not think you actually stated whether your people got into balance or not on 10 mEq sodium intake, and if not, what you accepted as a near-enough level. When we are doing routine studies, we sometimes, for practical reasons, have to accept 15 mEq/day in the urine as being near enough, but the state of sodium balance may be more critical here. Could I ask both speakers, what they would think would be the best criterion of determining that state of responsiveness – should it be sodium balance? Should it be, say, achieving balance at a certain intake over so many days? Or, is it really adequate to use diuretics as a stimulus? Would Dr. MULROW have any information on this point? I mean, could we get a reproducible effect from using diuretics, because clearly, this is a lot easier than running a sodium balance study.

MULROW: This is complicated because it depends what questions you are asking. We were asking different questions from those someone might be asking when looking for a diagnostic or therapeutic use of this kind of drug. We know from a lot of experience that after about 5 days low-sodium diet, a normal subject will be excreting certainly under 20 mEq of sodium a day, usually under 15, and the majority will be somewhere around 10 mEq/day.

Secondly, looking at weight as an index of volume depletion, we could see no correlation between the amount of weight loss and the hypotensive response to P113. Now, we only had 6 patients, and only 3 of them really had a striking effect, so our small amount of data would not substantiate the clear results that Dr. BRUNNER showed with respect to the amount of sodium lost, and the responsiveness to P113.

STOKES: With respect to what you said about 5 days at 10 mEq/day intake achieving balance: certainly, in Australia with the kind of high sodium diets that some people seem to eat here, you could not rely on that. Some people require as much as 8 or even 10 days to come into balance at 10, depending upon where they start. I asked the question to try and define what would be the minimum kind of regimen we could accept.

MULROW: Are these young subjects? Active subjects? Most of the ones we studied were young, active subjects, and usually they had a urinary sodium under 15 mEq/day, most of them under 10 mEq/day after 5 days.

BRUNNER: I would just say that we are still at the stage where we want to learn, and not where we are applying something that is well established. I believe that we have to do balance studies, because we know there is some tremendous individual variability of sodium loss in response to dietary restriction and diuretics. Even in normals, some lose much less than others. When we go to the different hypertensive states, this variability is much greater and I think it is not enough to know the urinary sodium on the 5th day, because some who are down to 10 mEq have lost very little salt, and others may have lost a lot.

Changes of Blood Pressure, Plasma Renin Activity and Plasma Aldosterone Concentration following the Infusion of Sar^1-Ile^8-Angiotensin II in Hypertensive, Fluid and Electrolyte Disorders

Toshihide Yamamoto, Kei Doi, Toshio Ogihara, Kiyoshi Ichihara, Takeshi Hata and Yuichi Kumahara

The Center for Adult Diseases; The Central Laboratory for Clinical Investigation, Osaka University Hospital, and Yodogawa Christ Hospital, Osaka

Contents

Introduction	174
Materials and Methods	176
Results	177
Normal Subjects in Three Different States of Sodium Balance	177
Patients with Hypertension, Fluid and Electrolyte Disorders	178
Relation between BP Changes and Pre-Infusion PRA	183
Relation between PAC Changes and Pre-Infusion PRA	183
Relationship between Changes in MBP, PRA and PAC	183
Discussion	183
Summary	186
References	187

Introduction

Angiotensin II blockade *in vivo* following intravenous infusion of sarcosine^1-alanine^8-angiotensin II (Sar^1-Ala^8-AII) in most hyperreninaemic, hypertensive subjects has been demonstrated to cause a reduction in blood pressure. This was interpreted to be evidence for the active involvement of angiotensin II (AII) in the pathogenesis of human hypertension associated with elevated plasma renin activity (PRA) [1]. Preliminary studies with another compound, sarcosine^1-isoleucine^8-angiotensin II (Sar^1-Ile^8-AII) revealed that some subjects with elevated PRA failed to show a reduction of BP and that subjects with low PRA showed elevation of BP [2, 3]; subsequently, it was reported that Sar^1-Ala^8-AII also has agonistic (pressor) activity [4]. The BP-reducing effect of Sar^1-Ala^8-AII in hypertensive subjects with high PRA can be augmented by prior administration of diuretics or by sodium restriction [4–6].

Angiotensin II analogues were originally thought to have antagonistic activity on aldosterone secretion [7]. It was later demonstrated that both Sar^1-Ala^8-AII and Sar^1-Ile^8-AII had agonistic or antagonistic activity for aldosterone secretion in experimental animals, depending on the dosage employed [8–10]. Elevation of PRA has been reported to occur concomitantly with reduction of BP during the infusion of angiotensin II analogues in man [1, 4] and in experimental animals [7, 9–12]. Thus, it seemed possible that aldosterone was stimulated by augmented secretion of renin as well as by intrinsic agonistic activity of the analogues [9]. The elevation of PRA following the infusion of angiotensin II analogues was attributed to antagonism of the intrarenal regulation of renin release by AII [4, 7, 9, 12]. In addition, 8-alanine angiotensin II, as well as AII, was shown to stimulate catecholamine secretion *in vivo* in dogs, and *in vitro* in adrenal medullary cell suspensions [13].

In planning clinical studies and therapeutic applications for any angiotensin II analogue, it has become necessary to obtain information about the antagonistic and agonistic activities of the compound *in vivo* regarding pressor activity, renin release, aldosterone secretion and catecholamine release from the adrenal medulla, as well as about the influence of dosage and sodium intake on these activities, all of which might possibly effect changes in the haemodynamics and renin-angiotensin-aldosterone system of the recipient. In the present study, changes in BP, PRA and plasma aldosterone concentration (PAC), following the intravenous infusion of Sar^1-Ile^8-AII, were determined in normal subjects. The effects of analogue dosage, prior administration of diuretics and sodium intake were also investigated. Further experiments were carried out in hypertensive subjects and those with fluid and electrolyte disorders to examine the effects of the analogue on BP, PRA and PAC. Analyses were made (a) to detect any relationships between changes of these parameters during the infusion and the clinical diagnosis of the recipient, and (b) to define antagonistic and agonistic effects upon BP, renin release, and aldosterone secretion in conditions of low and high PRA.

Materials and Methods

Sar^1-Ile^8-AII was synthesized by solid-phase procedure by SAKAKIBARA et al. (Peptide Institute, Protein Research Foundation, Osaka) and was processed into an injectable form by Daiichi Pharmaceutical Co. Ltd. (Tokyo). Subjects were 5 volunteers, including 4 of the authors, 18 subjects with essential hypertension, 7 with primary aldosteronism, 2 with Cushing's syndrome, 9 with renovascular hypertension, 6 with chronic renal failure, 5 with malignant hypertension, 1 with congestive heart failure, 2 with nephrotic syndrome, 8 with cirrhosis and ascites, 2 with Bartter's syndrome, 1 with renal tubular acidosis and 5 with phaeochromocytoma. Their diagnoses were made by clinical examination, routine laboratory studies and measurements of PRA and PAC. Diagnosis of renovascular hypertension was based on the demonstration of vascular lesions by renal arteriography. Diagnosis of Bartter's syndrome was made according to the criteria proposed by BARTTER et al. [14]. Diagnosis of renal tubular acidosis (proximal type) was confirmed by measurement of tubular reabsorption maximum of bicarbonate and pH changes following ammonium chloride administration. Diagnoses of all subjetcs with primary aldosteronism and phaeochromocytoma were confirmed by surgery.

Subjects with a variety of hypertensive and body fluid disorders were tested on unrestricted diet unless specified. Medications were discontinued in most cases at least one week prior to the infusion. Subjects with congestive heart failure, nephrotic syndrome and most cases of cirrhosis with ascites were kept on their medical treatment. After being in the supine position for 1 h, they were given i.v. infusion of normal saline at a rate of 1 ml/min for 30 min (control I). Then, saline was switched to Sar^1-Ile^8-AII diluted to a concentration of 200 ng/kg of body weight/ml with saline, which was given at a rate of 1 ml/min for 30 min (test). Then they were given saline for another 30 min (control II). In 5 normal subjects, the infusion was carried out at two dosage levels, i.e. 100 and 300 ng/kg weight/min. At the latter dose level, they were tested three times: without dietary sodium restriction, 4 h after the administration of 40 mg of furosemide and after being on a high salt intake (unrestricted diet supplemented with 6.6 g of NaCl in 3 divided doses) for 3 days. Most hypertensive subjects and those with fluid and electrolyte disorders received the infusion of Sar^1-Ile^8-AII at a rate of 200 ng/kg/min. Throughout the three phases of the infusion, BP was measured every 5 min over the brachial artery with a sphygmomanometer. Mean blood pressure (MBP) [1/3 ×

(systolic BP – diastolic BP) + diastolic BP] was calculated for each measurement of BP, and changes of BP were expressed as a percentage of the average of the 6 values for MBP in the test phase to that in the control I phase. Blood samples were drawn before the start and after the completion of Sar^1-Ile^8-AII infusion. Blood was collected into disodium edetate (1 mg/ml) and chilled immediately. Plasma was separated and stored at –20 °C until the time of assay. PRA was measured by a radioimmunoassay test kit based on the method originally described by HABER et al. [15] (Renin RIA kit, Dainabot laboratories, Tokyo), and PAC was measured, after extraction with dichlormethane, using a commercially available radioimmunoassay test kit (Aldok, CEA-IRE-Sorin, Paris).

Results

Normal Subjects in Three Different States of Sodium Balance

The i.v. infusion of Sar^1-Ile^8-AII at a rate of 100 ng/kg/min did not cause significant elevation of BP in normal subjects on an unrestricted salt intake, but the infusion at a rate of 300 ng/kg/min caused a significant elevation of BP, 10 mm Hg systolic and 20 mm Hg diastolic on the aver-

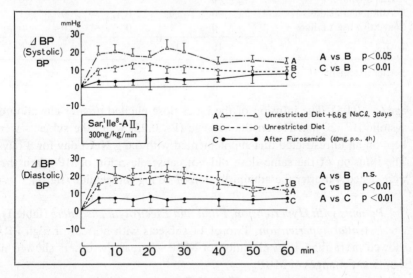

Fig. 1. Effect of diuretics and sodium intake on blood pressure response to Sar^1-Ile^8-AII in normal subjects.

Table I. Changes of mean blood pressure following Sar¹-Ile⁸-AII infusions in subjects with hypertensive, fluid and electrolyte disorders

	Number of cases	Percent change of mean BP				
		$\infty+20$	$\infty+10$	$+10\infty-10$	-10∞	-20∞
Hypertensive group						
Essential hypertension						
High PRA	2	0	0	2	0	0
Normal PRA	9	1	1	7	0	0
Low PRA	7	2	3	2	0	0
Primary aldosteronism	7	0	4	3	0	0
Cushing's syndrome	2	1	1	0	0	0
Renovascular hypertension	9	0	1	6	2	0
Chronic renal failure	6	1	3	2	0	0
Malignant hypertension	5	0	0	4	0	1
Phaeochromocytoma						
Hypertensive phase	3	3	0	0	0	0
Normotensive phase	2	0	0	2	0	0
Nephrotic syndrome with hypertension	2	2	0	0	0	0
Normotensive group						
Liver cirrhosis with ascites	8	0	0	6	0	2
Bartter's syndrome	2	0	0	1[a]	1	0
Renal tubular acidosis	1	0	0	0	0	1
Congestive heart failure	1	0	1	0	0	0
Total	66	10	14	35	3	4

[a] After treatment.

age ($p<0.05$). The infusion of the latter dose elicited greater elevation of systolic BP, 20 mm Hg on the average ($P<0.05$) when the subjects were kept on an unrestricted diet supplemented with 6.6 g NaCl/day for 3 days. The infusion of the same dose did not cause elevation of BP when they were tested 4 h after the administration of 40 mg furosemide (fig. 1).

Patients with Hypertension, Fluid and Electrolyte Disorders (table I)

Essential hypertension. Two of 11 subjects with normal or high PRA showed more than 10% elevation of MBP whereas the others showed no significant changes of MBP.

Low-renin hypertension. Five of 7 subjects with low-renin essential hypertension, 4 of 7 subjects with primary aldosteronism and 2 subjects

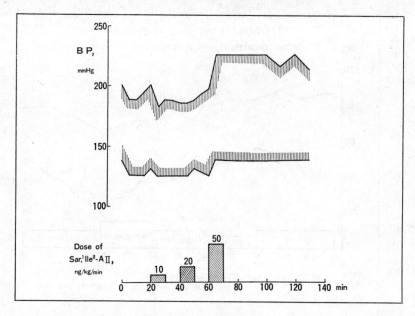

Fig. 2. Blood pressure to Sar[1]-Ile[8]-AII infusion in patient with primary aldosteronism.

with Cushing's syndrome showed more than 10% elevation of MBP. In particular, one subject with Cushing's syndrome showed marked (66%) elevation of MBP. Three subjects with primary aldosteronism were given infusions of Sar[1]-Ile[8]-AII at three dosage levels, namely 10, 20 and 50 ng/kg/min (each dose for 10 min) while they were kept on an unrestricted diet. The elevation of BP was not evident with smaller doses, but the subjects showed significant elevations of BP, lasting more than 60 min after discontinuation of the infusion, in response to 50 ng/kg/min (fig. 2).

Renovascular hypertension. All but one subject with renovascular hypertension had high PRA. A reduction of MBP by more than 10% was observed in only 2 of 9 subjects, while in one subject MBP rose by 13%.

Malignant hypertension. Of 5 subjects with malignant hypertension, only one, whose pre-infusion PRA was 8.0 ng/ml/h, showed a reduction of MBP (20%). However, in another with a PRA of 17 ng/ml/h, BP was not reduced.

Hypertension due to chronic renal failure. Four of 6 subjects with chronic renal failure and hypertension showed more than 10% elevation of MBP.

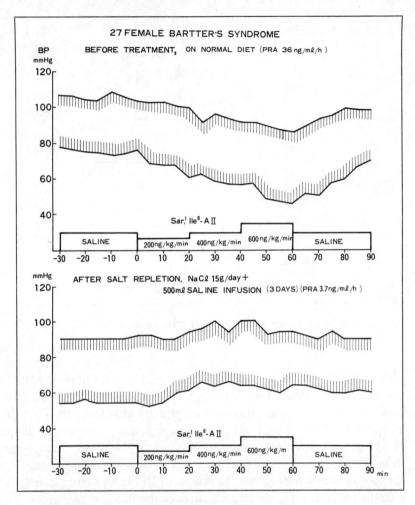

Fig. 3. Blood pressure response to Sar¹-Ile⁸-AII in patient with Bartter's syndrome.

Oedematous states. In two subjects with nephrotic syndrome accompanied by hypertension, MBP rose by more than 20%. A similar rise occurred in one normotensive subject with congestive heart failure (arteriosclerotic aetiology). Two of 8 subjects with liver cirrhosis with ascites showed more than 20% reduction of MBP and the rest had no significant changes of BP.

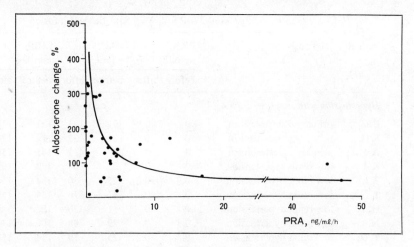

Fig. 4. Relation between plasma renin activity and change of mean blood pressure.

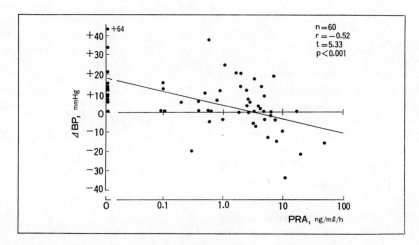

Fig. 5. Relation between plasma renin activity and aldosterone change.

Bartter's syndrome and renal tubular acidosis. One subject with Bartter's syndrome and one with renal tubular acidosis showed a reduction of BP, the degree of which appeared to be dose-dependent (fig. 3). However, the former subject, tested 3 days after 15 g NaCl supplementation of regular hospital diet coupled with 500 ml saline infusion, and another subject with the same condition, chronically treated with salt and potassium sup-

Table II. Changes of PRA, PAC and MBP following Sar[1]-Ile[8]-AII infusion

Patient	Diagnosis	PRA, ng/ml/h		PAC, pg/ml		MBP, mm Hg	
		before	after	before	after	before	after
Hypertension with suppressed PRA							
I.K.	primary aldosteronism	0	0	51	103	123	138
S.H.	primary aldosteronism	0	0	121	233	112	126
A.H.	primary aldosteronism	0	0	329	423	141	154
K.I.	low renin hypertension	0	0	23	88	129	155
T.I.	low renin hypertension	0.3	0.2	25	30	134	150
T.N.	low renin hypertension	0	0	141	216	134	150
H.T.	low renin hypertension	0.1	0	51	136	120	139
N.M.	low renin hypertension	0.3	0	29	88	100	134
T.H.	low renin hypertension	0.3	0.2	25	30	134	157
Normal control on high sodium intake							
T.Y.		0.8	1.0	59	116	86	110
T.O.		1.5	1.1	49	71	83	107
K.I.		1.3	0.7	12	38	91	104
K.Iw.		0.9	1.1	44	59	86	99
T.H.		1.0	0.6	22	73	80	94
Hypertension with high PRA							
H.Y.	essential hypertension	5.0	6.6	84	47	137	154
A.N.	renovascular hypertension	7.3	4.5	174	174	131	154
Y.K.	renovascular hypertension	7.9	5.9	148	188	123	146
N.F.	malignant hypertension	17.0	10.0	564	225	169	168
K.A.	renovascular hypertension	4.1	1.7	112	141	146	150
K.O.	essential hypertension	4.5	4.1	114	24	135	133
Normotension with high PRA							
H.M.	Bartter's syndrome	36	70	1100	520	78	64
M.I.	renal tubular acidosis	20	60	44	24	70	55
M.M.	liver cirrhosis with ascites	3.3	2.8	52	28	105	99
O.K.	liver cirrhosis with ascites	31.5	50	564	249	72	69

plementation, did not show reduction of BP following the infusion of Sar[1]-Ile[8]-AII.

Phaeochromocytoma. Of 5 subjects with phaeochromocytoma, three, who were hypertensive at the time of the infusion, were found to have a rise in BP exceeding 10%. In the remaining two patients, who were normotensive at the time of the infusion, BP did not change. PRA in the first three subjects was 4.4, 0.9 and 2.5 ng/ml/h, and in the last two, 6.4 and 3.6 ng/ml/h.

Relation between BP Changes and Pre-Infusion PRA (fig. 4)

Out of 66 subjects who underwent the infusion sequence, PRA was measured in 45 both before and after the infusion. In most of these subjects PAC was also measured. Subjects with low PRA tended to exhibit elevations of MBP and those with high PRA to show reduction of MBP following the Sar^1-Ile^8-AII infusion. There was a negative correlation between MBP changes and pre-infusion PRA ($r = -0.42$, $p<0.001$).

Relation between PAC Changes and Pre-Infusion PRA (fig. 5)

A curvilinear relation was found between pre-infusion PRA and changes of PAC (post-infusion PAC expressed as a percentage of pre-infusion PAC) ($r = -0.52$, $p<0.01$).

Relationships between Changes in MBP, PRA and PAC (table II)

Data of MBP, PRA and PAC of subjects with either high or low PRA were extracted in order to analyse relationships between changes in these parameters. Hyporeninaemic subjects, both normotensive and hypertensive, showed significant elevation of MBP and PAC and reduction of PRA. In subjects with high PRA changes in MBP and PAC were generally in the opposite direction to those in PRA, though the changes were not statistically significant by paired *t*-test.

Discussion

Sar^1-Ile^8-AII was found to have agonistic activity resembling the pressor action of AII in normal subjects. This appears to correlate with the dosage of the analogue, and the sodium intake of the recipient. The findings suggest that the response is determined by the level of endogenous angiotensin II rather than sodium balance *per se*, the agonistic activity being inversely proportional to the former. This is analogous to the situation with infused AII, the pressor activity of which is altered by the level of endogenous AII and may reflect the degree of occupancy of AII receptors with endogenous AII [16].

Parallel shifts of dose-pressor response curves of AII occur following infusion of Sar^1-Ile^8-AII [3], indicating competitive antagonism of AII by the analogue. Thus, a receptor common to AII and its analogues may exist, possibly in the peripheral vascular smooth muscle, and the observed agonistic activity of Sar^1-Ile^8-AII is probably mediated via the same pressor receptor as for AII.

Only a few subjects with renovascular and malignant hypertension accompanied by high PRA showed reductions in BP following the infusion of Sar^1-Ile^8-AII. The present results differ from those obtained with Sar^1-Ala^8-AII, which has been shown to reduce BP in most subjects with renovascular hypertension and elevated PRA. Admittedly, the preparation, the dosage (in relation to potency) and the duration of infusion were different for each study. Dependency of angiotensin II blockade on sodium depletion has been emphasized in regard to other angiotensin II analogues [6, 17]. Daily sodium intake was estimated to be between 8 and 13 g/day for residents in the Osaka area where the present study was carried out [18]. We could have elicited more frequent positive responses had we routinely used sodium depletion prior to the administration of Sar^1-Ile^8-AII. Agonist responses have also been reported with Sar^1-Ala^8-AII [4, 19]. Therefore, knowledge of the relative agonist and antagonist potencies of these compounds in relation to a given level of sodium intake is required to plan clinical studies.

With respect to subjects with disorders characterised by fluid retention, those with chronic renal failure, congestive heart failure and nephrotic syndrome showed variable elevations of BP in response to Sar^1-Ile^8-AII, whereas some of those with liver cirrhosis and ascites showed significant reduction of BP. The findings in subjects with liver cirrhosis and ascites were comparable to those observed in dogs with thoracic caval constriction following the infusion of Sar^1-Ala^8-AII [20]. In the former group of subjects, the renal excretion of sodium was impaired so that, conceivably, the vascular volume tended to be expanded. In the latter group, however, the effective vascular volume was presumably reduced because of decreased colloid oncotic pressure. This appears to be at least one reason for the observed difference in response to Sar^1-Ile^8-AII in these two kinds of body fluid excess. Similarly, reduction of BP following Sar^1-Ile^8-AII in subjects with Bartter's syndrome and renal tubular acidosis (in which impaired renal reabsorption of sodium has been suggested) could be understood on the basis of reduced vascular volume [21–23]. This explanation was supported by the observation that one subject with Bartter's syndrome who was chronically treated with salt and potassium, and another who received an acute salt load, responded to Sar^1-Ile^8-AII infusion with a slight elevation of BP. Angiotensin II is probably participating in the maintenance of normal BP in those with deficient effective vascular volume.

In 3 of 5 subjects with phaeochromocytoma, hypertensive at the time of infusion, BP rose in response to Sar^1-Ile^8-AII. Their average PRA was high. It has been reported that most patients with phaeochromocytoma have a normal blood volume [24], so that expanded blood volume can hardly be an explanation for a pressor response to Sar^1-Ile^8-AII. Although urinary excretion of free noradrenaline and adrenaline were not increased in normal subjects following hypertension induced by Sar^1-Ile^8-AII [YAMAMOTO et al., unpublished data], catecholamines might have been released from the *tumour* in these subjects as suggested from a study in dogs by PEACH [13]. Of particular interest is an exaggerated hypertensive response seen in one subject with Cushing's syndrome. Hypertension in Cushing's syndrome was thought to be due to sodium retention attributed to mineralocorticoid excess [25]. Of late, enhanced vascular reactivity to pressor agents, coupled with increased angiotensin formation resulting from elevated renin substrate levels, has been proposed to explain hypertension in this condition [26]. Either of these mechanisms could be operative in exaggerating hypertensive responses to Sar^1-Ile^8-AII. We think that the administration of angiotensin II analogues should be discarded in these two conditions, i.e. phaeochromocytoma and Cushing's syndrome, because of a risk of cardiovascular accident due to excessive hypertensive responses.

A low degree of correlation was observed between pre-infusion PRA and changes of BP as well as between pre-infusion PRA and changes of PAC (fig. 4, 5). These correlations, however, were not obvious in the region of normal PRA. The directions of changes of PRA and PAC were quite variable in subjects with normal PRA. This was conceivably influenced by many variables as discussed in the foregoing paragraphs. However, relationships between changes in these parameters were less complicated in subjects with low and high PRA. Significant elevation of MBP and PAC and reduction of PRA were observed in subjects with low PRA, while changes of these parameters were generally in the opposite directions in normotensive subjects with high PRA. Hypertensive responses to Sar^1-Ile^8-AII in subjects with low PRA and hypotensive responses in normotensive subjects with high PRA could be understood on the basis of an excess or deficit of vascular volume. The failure of BP to fall in hypertensive subjects with high PRA indicates that their hypertension has a component attributable to excess vascular volume. The reciprocal changes of PRA (except in hypertensive subjects with high PRA) were consistent with another study [15]. AII-induced inhibition of renin release, thought to be mediated via an intrarenal mechanism [1, 12, 29], has been established [27, 28]. Sar^1-Ile^8-AII could possible be both antagonistic and agonistic for the release of renin, though PRA changes secondary to changes of BP could not be ruled out in the present study. It is of interest that PAC changed in different directions in the low and high renin groups. These changes were unlikely to be secondary to effects on PRA, for reduction of PAC was observed in subjects with high PRA, both normotensive and hypertensive. Rather, they appear to be due to agonistic and antagonistic activities of Sar^1-Ile^8-AII upon aldosterone secretion. On the basis of experiments in dogs using doses comparable to those used in the present study, it has been suggested that Sar^1-Ile^8-AII possesses agonistic activity for aldosterone secretion [10]. Excessive amounts of Sar^1-Ile^8-AII were required to antagonise aldosterone secretion induced by AII in dexamethazone-suppressed, nephrectomized dogs [8]. The present observation that Sar^1-Ile^8-AII has both antagonistic and agonistic activities at one dose level *in man* needs further study.

Summary

1-Sarcosine, 8-isoleucine angiotensin II (Sar^1-Ile^8-AII) was infused intravenously in 5 normal volunteers and 66 subjects with various hypertensive, fluid and electrolyte disorders. Changes of blood pressure (BP), plasma renin activity (PRA) and plasma aldosterone concentration (PAC) were studied. In normal subjects, Sar^1-Ile^8-AII showed pressor (agonistic) activity, which was related to both dosage and sodium intake. Hyporeninaemic hypertensive subjects (primary aldosteronism) showed pressor responses to a smaller dose of this compound than the dose employed in normal subjects. Hyporeninaemic hypertensive subjects and normal volunteers after 3 days of high sodium intake showed significant elevations of BP and PAC and reduction of PRA. Changes of BP, PAC and PRA in normoreninaemic subjects were variable. Hyperreninaemic normotensive subjects including those with Bartter's syndrome, renal tubular acidosis or liver cirrhosis with ascites showed reduction of BP and PAC and elevation of PRA. The results indicate that the compound has both agonistic and antagonistic activities for blood pressure; which of these is obtained apparently depends upon endogenous angiotensin II levels, as well as the dosage employed. The results in subjects with high and low PRA suggest that the compound has antagonist and agonist actions at 3 sites of angiotensin II action, i.e. peripheral vascular bed, renin release mechanism from juxta-glomerular apparatus and the zona glomerulosa of the adrenals.

References

1 BRUNNER, H. R.; GAVRAS, H.; LARAGH, J. H., and KEENAN, R.: Angiotensin II blockade in man by sar[1]-ala[8]-angiotensin II for understanding and treatment of high blood pressure. Lancet *ii:* 1045–1047 (1973).
2 OGIHARA, T.; YAMAMOTO, T., and KUMAHARA, Y.: Angiotensin II blockade (letter). Lancet *i:* 219 (1974).
3 OGIHARA, T.; YAMAMOTO, T., and KUMAHARA, Y.: Clinical application of synthetic angiotensin II analogue. Jap. Circulation J. *38:* 997–1007 (1974).
4 ANDERSON, G. H.; FREIBERG, J. M.; DALAKOS, T. G., and STREETEN, D. H. P.: Effect of angiotensin sensitivity on renin release. The Endocrine Society Program, 57th Ann. Meet., New York 1975, abstr., p. 139.
5 DONKER, A. J. M. and LEENAN, F. H. H.: Infusion of angiotensin II analogue in two patients with unilateral renovascular hypertension. Lancet *ii:* 1535–1537 (1974).
6 STREETEN, D. H. P.; ANDERSON, G. H.; FREIBERG, J. M., and DALAKOS, T. G.: Use of an angiotensin II antagonist (saralasin) in the recognition of 'angiotensinogenic' hypertension. New Engl. J. Med. *292:* 657–662 (1975).
7 JOHNSON, J. A. and DAVIS, J. O.: Effect of a specific competitive antagonist of angiotensin on arterial pressure and adrenal steroid secretion in dogs. Circulation Res. *32–33:* suppl. I, pp. 159–168 (1973).
8 BRAVO, E. L.; KHOSLA, M. C., and BUMPUS, F. M.: Vascular and adrenocortical responses to a specific antagonist of angiotensin II. Am. J. Physiol. *228:* 110–114 (1975).
9 STEELE, J. M. and LOWENSTEIN, J.: Differential effects of an angiotensin II analogue on pressor and adrenal receptors in the rabbit. Circulation Res. *35:* 592–600 (1974).
10 BECKENHOFFER, R.; UHLSCHMID, G.; VETTER, W.; ARMBRUSTER, H.; NUSSBERGER, J.; SCHMID, U., and SIEGENTHALER, W.: The effect of the angiotensin analogue 1-sar-8-ile angiotensin and renin secretion in dogs (abstr.). Acta endocr., Copenh. *78:* suppl. 193, p. 134 (1975).
11 GOODFRIEND, T. L. and PEACH, M. L.: Angiotensin III: (des-aspartic acid[1])-angiotensin II, evidence and speculation for its role as an important agonist in the renin angiotensin system. Circulation Res. *36–37:* suppl. I, pp. 38–48 (1975).

12 FREEMAN, R. H.; DAVIS, J. O., and LOHMEIER, T. E.: Des-1-asp-angiotensin II, possible intrarenal role in homeostasis in the dog. Circulation Res. *37:* 30–34 (1975).
13 PEACH, M. J.: Adrenal medullary stimulation induced by angiotensin II and analogues. Circulation Res. *28–29:* suppl. 2, pp. 107–117 (1971).
14 BARTTER, F. C.; PROVONE, P.; GILL, J. R., and MACCARDLE, R. C.: Hyperplasia of the juxtaglomerular complex with hyperaldosteronism and hypokalemic alkalosis. Am. J. Med. *33:* 811–828 (1962).
15 HABER, E.; KOERNER, T.; PAGE, L. B.; KLIMAN, B., and PURNODE, A.: Application of a radioimmunoassay for angiotensin I to the physiologic measurements of plasma renin activity in normal human subjects. J. clin. Endocr. Metab. *29:* 1349–1355 (1969).
16 THURSTON, H. and LARAGH, J. B.: Prior receptor occupancy as a determinant of the pressor activity of infused angiotensin II in the rat. Circulation Res. *36:* 113–117 (1975).
17 BRUNNER, H. R.; GAVRAS, H.; LARAGH, J. B., and KEENAN, R.: Hypertension in man, exposure of the renin and sodium components using angiotensin II blockade. Circulation Res. *34:* suppl. I, pp. 35–43 (1974).
18 KOMACHI, Y.: Difference of hypertension and arteriosclerosis by district and by occupation. Jap. Circulation J. *33:* 1473–1476 (1969).
19 MARKS, L. S.; MAXWELL, M. H., and KAUFMAN, J. J.: Saralasin bolus test. Rapid screening procedure for renin-mediated hypertension. Lancet *ii:* 784–787 (1975).
20 JOHNSON, J. A. and DAVIS, J. O.: Effects of a specific competitive antagonist of angiotensin II arterial pressure and adrenal steroid secretion in dogs. Circulation Res. *32–33:* suppl. I, pp. 159–163 (1973).
21 CANNON, P. J.; LEEMING, J. M.; SOMMERS, S. C.; WINTERS, R. W., and LARAGH, J. B.: Juxtaglomerular cell hyperplasia and secondary hyperaldosteronism (Bartter's syndrome). A re-evaluation of the pathophysiology. Medicine *47:* 107–131 (1968).
22 GOODMAN, A. D.; VAGNUCCI, A. H., and HARTROFT, P. M.: Pathogenesis of Bartter's syndrome. New Engl. J. Med. *281:* 1435–1439 (1969).
23 FLEISHMAN, S. J.; SENIOR, B., and SUZMAN, M. M.: Renal tubular acidosis, the role of defective renal tubular sodium reabsorption and secondary hyperaldosteronism in its pathogenesis. Archs intern. Med. *104:* 613–618 (1959).
24 TARAZI, R. C.; DUSTAN, H. P.; FROHLICH, E. D.; GIFFORD, R. W., and HOFFMAN, G. C.: Plasma volume and chronic hypertension. Relationship to arterial pressure levels in different hypertensive diseases. Archs intern. Med. *125:* 835–842 (1970).
25 KAPLAN, N. N.: Adrenal causes of hypertension. Archs intern. Med. *133:* 1001–1004 (1975).
26 KRAKOFF, L.; NICOLIS, G., and AMSEL, B.: Pathogenesis of hypertension in Cushing's syndrome. Am. J. Med. *58:* 216–220 (1975).
27 BUNAG, R. D.; PAGE, I. H., and McCUBBIN, J. W.: Inhibition of renin release by vasopressin and angiotensin. Cardiovasc. Res. *1:* 67–73 (1967).
28 BLAIR-WEST, J. R.; COGHLAN, J. P.; DENTON, D. A.; FUNDER, J. W.; SCOGGINS,

B. A., and WRIGHT, R. D.: Inhibition of renin secretion by systemic and intrarenal angiotensin infusion. Am. J. Physiol. *220:* 1309–1315 (1971).

29 VANDONGEN, R.; PEART, W. S., and BOYD, G. W.: Effect of angiotensin II and its nonpressor derivatives on renin secretion. Am. J. Physiol. *226:* 227–282 (1974).

T. YAMAMOTO, MD, The Center for Adult Diseases, Osaka, 1-chome, Nakamichi, Higashinari-ku, *Osaka* (Japan)

Angiotensin II Blockade in Hypertensive Dialysis Patients

G. A. MacGregor and P. M. Dawes

Department of Medicine, Charing Cross Hospital Medical School, London and ICI, Alderley Park, Cheshire

Contents

Introduction	190
Methods	191
Case Reports	192
Discussion	196
Summary	198
Acknowledgements	198
References	198

Introduction

Patients with chronic renal failure on maintenance haemodialysis may have high blood pressure. Two mechanisms seem to be involved [1–5]. First, there may be sodium and water overload with a relative suppression of angiotensin II, when the blood pressure can be controlled by the removal of sodium and water (i.e. weight) at dialysis. Second, there may be hypertension associated with an abnormally high level of angiotensin II. In these patients, a rapid removal of weight may cause the arterial pressure to rise, and subsequently the blood pressure is difficult to control. In practice, it is not always easy to distinguish these two mechanisms within a group of hypertensive dialysis patients even when measurements of angiotensin II or plasma renin activity are readily available, particularly as in many of the dialysis patients both mechanisms may be responsible for the hypertension.

Sar[1]-Ala[8]-angiotensin II (saralasin, P113), a competitive inhibitor of angiotensin II, has been shown to lower blood pressure in hypertensive patients in whom angiotensin II is causing the increased blood pressure [6, 7].

We report here preliminary studies on 5 dialysis patients with high blood pressure who were infused with saralasin. The patients were selected because they illustrate the different mechanisms that may be involved in raising the blood pressure of dialysis patients.

Methods

Five patients consented to take part in the study, the purpose of which was fully explained to them. They were being dialysed twice a week for 10 h per dialysis, using a Meltec multipoint I square metre kidney. Patients were infused with saralasin on the morning of a normal dialysis day (predialysis). In two, the infusion was repeated on the day after dialysis (postdialysis) at least 12 h after the dialysis had finished. An intravenous cannula was inserted. After a control period of 45–60 min, saralasin was infused at rates of 0.25, 1.0, 5.0, 10.0 μg/kg/min, each infusion rate being given for 15 min. Different dilutions of saralasin in 5% dextrose were used, so that the pump speed was kept constant at 0.76 ml/min. Blood pressure was measured every 30 sec to 2 min using an Arteriosonde 1217 (Roche), a semi-automatic ultrasound sphygmomanometer [8], and measurements were continued for 1 h after saralasin was stopped. Standing blood pressure was measured approximately every 15 min, after standing upright for 1 min and then two to four times at 1-min intervals. Blood was taken before infusion for measurement of angiotensin II and plasma renin activity (PRA), and for PRA just before saralasin infusion was stopped. No patient had more than 50 ml of blood removed or more than 50 ml of 5% dextrose given during the infusion.

Plasma angiotensin II was measured by the method of DUSTERDIECK and MCELWEE [9], and PRA was measured by radioimmunoassay of angiotensin I generated during plasma incubation for 90 min at 37 °C at pH 6.0 in presence of EDTA, 8-hydroxyquinoline and phenylmethylsulphonylfluoride. Mean values of angiotensin II and PRA for these assays in 17 normals on their normal diet (mean urinary sodium excretion

148 mEq/24 h), measured sitting upright between 10 a.m. and 12 midday, were 28.4 pg/ml (range 18.9–45) and 5.1 ng/ml/h (range 1.9–8.9), respectively. The same 17 normals were sodium-deprived by a low salt diet (10 mEq/day) and on the fifth day of this diet mean angiotensin II and PRA were 63.1 pg/ml (range 35–100) and 17.4 ng/ml/h (range 8–25) with a mean urinary sodium excretion of 11.0 mEq/24 h.

Case Reports

Patient No. 1. A 26-year-old man with a five-year history of progressive renal failure and mild hypertension, assumed to be due to glomerulonephritis, was transferred for haemodialysis in September 1973. In October 1973, he had a renal transplant but developed severe hypertension, requiring a bilateral nephrectomy. One month later, the transplant failed and was removed. Subsequently, during maintenance dialysis, his blood pressure was mildly raised (150–160 mm Hg systolic, 90–100 mm Hg diastolic), especially when he was fluid-overloaded. The blood pressure was easily controlled with the removal of sodium and water during dialysis. At a time when his weight was 3 kg higher than after his previous dialysis, his jugular venous pressure was +6 cm, and his lying blood pressure 140/90 mm Hg, saralasin infusion did not cause any fall in either lying or standing blood pressure. There was, however, a mild transient rise in blood pressure at the beginning of the infusion of saralasin, lasting 3–5 min, which also occurred when the infusion rate was increased. Angiotensin II and PRA prior to infusion were low at 12.9 pg/ml and 0.27 ng/ml/h, respectively. During saralasin infusion, PRA was 0.32 ng/ml/h. Subsequently, blood pressure was controlled by a 1.5-kg reduction in pre-dialysis weight.

The high blood pressure in this bilaterally nephrectomised patient was volume-dependent, in that he had very low values of angiotensin II and PRA before dialysis, and no fall in blood pressure with saralasin. In addition, the blood pressure fell with weight removal during dialysis and eventually became controllable. The transient rise in blood pressure at the beginning of infusion, and on increasing the dose of saralasin, has previously been described in sodium-loaded normals and low renin hypertensives, both situations where angiotensin II is low [10, 11].

Patient No. 2. A 22-year-old man was first seen in 1969 with nephrotic syndrome and normal blood pressure. Renal biopsy showed focal and segmental glomerulonephritis. Over the next five years, renal function gradually deteriorated and the blood pressure rose. The blood pressure was controlled with methyldopa. Haemodialysis was started in March 1975 when hypotensive therapy was stopped. Blood pressure was in the range 150–160/100–110 mm Hg. It fell during and after dialysis with removal of sodium and water. Pre-dialysis, when his blood pressure was 160/110 mm Hg and his weight was 2 kg above his normal post-dialysis weight, a saralasin infusion caused no change in lying or standing blood pressure. Immediately before the infusion of saralasin, angiotensin II and PRA were in the low normal range at 18.4 pg/ml and 4.64 ng/ml/h, respectively. Subsequently, with a reduction of 2 kg in his pre-dialysis weight, blood pressure has become normal.

The high blood pressure in this patient also seems to have been volume-dependent in that pre-dialysis he had low normal values of angiotensin II and PRA, and no fall in blood pressure with saralasin. In addition, the blood pressure fell with sodium and water removal during dialysis, and later his high blood pressure was controlled with a gradual reduction in his pre-dialysis weight.

Patient No. 3. A 32-year-old man, with a six-month history of hypertension and chronic renal failure thought to be due to glomerulonephritis, continued to deteriorate in spite of control of his blood pressure with methyldopa. He was transferred for dialysis in November 1974. Methyldopa was stopped. Over the next 3 months, his pre-dialysis weight was reduced by 5 kg in an attempt to control his blood pressure. Blood pressure prior to each dialysis, however, was around 180/115 mm Hg. Saralasin was infused on the morning of his usual dialysis day, when the lying blood pressure was 210/119 mm Hg. He had gained 1 kg since his last dialysis. Plasma angiotensin II and PRA prior to infusion were extremely high at 493 pg/ml and 69.1 ng/ml/h, respectively. Saralasin infusion reduced lying and standing blood pressure from 210/119 and 215/118, respectively, to 157/83 mm Hg and 135/75 mm Hg. The infusion was stopped at an infusion rate of 1 μg/kg/min because of the risk of hypotension developing at greater infusion rates. During the saralasin infusion, PRA rose to 151.6 ng/ml/h. After the saralasin was stopped, the blood pressure returned towards pre-infusion values within 30 min. After dialysis, when 1 kg of weight had been removed, lying blood pressure remained elevated at 185/121 mm Hg but fell to 120/70 mm Hg on standing. The infusion of saralasin therefore, was not repeated post-dialysis. Subsequently, his pre-dialysis weight was allowed to rise by 2 kg and blood pressure has been well controlled with propranolol 80 mg t.d.s. and hydrallazine 25 mg t.d.s.

The high blood pressure in this patient was angiotensin II-dependent as evidenced by: very high levels of angiotensin II and PRA pre-dialysis, the fall of blood pressure to normal with saralasin predialysis, and the development of postural hypotension with further weight removal during dialysis.

Patient No. 4. A 26-year-old man with a six-month history of renal failure thought to be due to glomerulonephritis had his high blood pressure controlled with methyldopa and propranolol. Renal function continued to deteriorate and haemodialysis was begun in December 1974. Hypotensive therapy was stopped at this time and his blood pressure was controlled to some extent with sodium and water removal during dialysis. Over the next few months, his pre-dialysis weight was reduced by 15 kg with transient improvement in his blood pressure. The blood pressure, however, was not satisfactorily controlled and it was noted that his blood pressure tended to be higher the day after dialysis when he had lost 2–3 kg in weight. On April 16th 1975, on the morning before dialysis, lying blood pressure was 175/104 mm Hg; his weight was 3 kg above his normal post-dialysis weight; jugular venous pressure was raised by 5 cm, with no oedema. PRA prior to saralasin was in the low normal range, at 1.98 ng/ml/h. Saralasin infusion caused no change in lying or standing blood pressure (fig. 1) and PRA did not change during saralasin infusion, being 2.24 ng/ml/h. 12 h after dialysis, during which 3 kg of weight was removed, lying blood pressure was higher at 174/120 mm Hg than pre-dialysis and PRA had risen above the normal range to 24.3 ng/ml/h. Saralasin infusion at this time caused the lying blood pressure to fall to 153/86 mm Hg (fig. 1). During the infusion, PRA

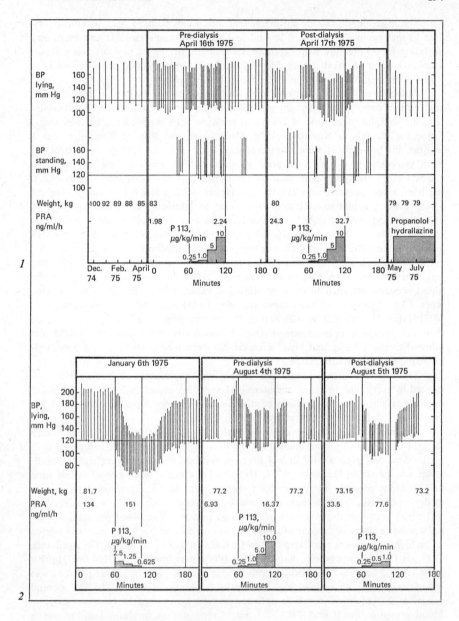

Fig. 1. Patient No. 4. Blood pressure changes over 9 months and with saralasin (P113) infusion pre- and post-dialysis.

Fig. 2. Patient No. 5. Blood pressure changes with saralasin (P113) infusion when first seen and after 8 months dialysis pre- and post-dialysis.

rose further to 32.7 ng/ml/h. Subsequently, no further attempt was made to control his blood pressure with weight removal; instead, his predialysis weight has been kept constant and his blood pressure has been well controlled with propranolol 40 mg t.d.s. and hydrallazine 25 mg t.d.s.

Pre-dialysis, the high blood pressure in this patient appears to have been volume-dependent, whereas post-dialysis, after the removal of 3 kg of sodium and water, the blood pressure appears to have become angiotensin II and volume-dependent. Pre-dialysis, there was a low value of PRA and no fall in blood pressure with saralasin. Post-dialysis, with the reduction in weight, PRA had risen but was not above the range of salt-deprived normal individuals. At the same time, the blood pressure also had risen and saralasin now caused his blood pressure to fall, but not to normal levels.

Patient No. 5. A 30-year-old woman was seen with malignant hypertension in October 1974. Blood pressure was 250/170 mm Hg and there was papilloedema. Blood pressure was very difficult to control and acute oliguric renal failure developed. The patient was transferred for dialysis on January 5th 1975. At this time, she was in mild left ventricular failure with a blood pressure of 207/122 mm Hg lying. On January 6th, a saralasin infusion caused the lying blood pressure to fall to 131/66 mm Hg (fig. 2). At the same time, there was marked improvement in symptoms and in the signs of left ventricular failure. Standing blood pressure was not measured in view of the risk of hypotension. Plasma angiotensin II and PRA prior to infusion were very high at 400 pg/ml and 134 ng/ml/h, and during the infusion the PRA rose to 151 ng/ml/h. Haemodialysis was started 1 h after stopping saralasin when the blood urea was 350 mg/100 ml and plasma creatinine was 15.0 mg/100 ml. The blood pressure was initially controlled with a combination of diazoxide, propranolol and hydrallazine but, in spite of hypotensive therapy and some weight removal at dialysis, the blood pressure remained difficult to control. This was aggravated by the large amounts of fluid which she drank so that her weight between dialysis might rise by up to 6 kg. Eight months later, 3 days after stopping hypotensive drugs, a second saralasin infusion was given on the morning of a normal dialysis day. At this time, her weight was 5 kg above her normal post-dialysis weight, and the jugular venous pressure was raised by 8 cm: there were bilateral basal crepitations; PRA was 6.93 ng/ml/h. Her lying blood pressure was 196/121 mm Hg. Saralasin infusion reduced her lying blood pressure to 165/106 mm Hg (fig. 2) and PRA rose during infusion to 16.37 ng/ml/h. 12 h after dialysis, when 4 kg of weight had been removed, PRA had risen to 33.5 ng/ml/h, and lying blood pressure was 189/122 mm Hg. Saralasin infused only up to a rate of 1 μg/kg/min reduced her lying blood pressure to 146/100 mm Hg (fig. 2) and PRA rose during infusion to 77.6 ng/ml/h. The infusion rate of saralasin was not increased because she had recently clotted her arterio-venous shunt, and hypotension might have caused her shunt to reclot. Subsequently, this patient has managed to restrict her fluid intake, and her blood pressure is well controlled by keeping her weight steady and treating her with propranolol 40 mg t.d.s. and hydrallazine 25 mg t.d.s.

Before being placed on maintenance haemodialysis, this patient had angiotensin II-dependent accelerated hypertension. Subsequently, when she was on dialysis and became fluid-overloaded, the hypertension pre-dialysis appeared to become partly volume and partly angiotensin II-dependent. PRA at this time was around the upper limit of the normal subjects on an ordinary diet, but in relation to the gross volume overload such a concentration was inappropriately high. Post-dialysis, with a 4-kg reduction in weight, there was a rise in PRA above the range of sodium-deprived normal subjects, and the blood pressure now appeared to become even more angiotensin II-dependent.

Discussion

This preliminary study suggests that angiotensin II blockade with saralasin separates hypertensive maintenance haemodialysis patients into two distinct groups. Firstly, there is a group in whom saralasin infusion does not lower blood pressure. The high blood pressure in this group falls with weight removal during dialysis, and with a gradual reduction in pre-dialysis weight these patients became normotensive. Secondly, there is a smaller group in whom saralasin infusion does lower blood pressure either pre- or post-dialysis. The blood pressure in this group does not fall with weight removal during dialysis, or with a gradual reduction in pre-dialysis weight. These patients require anti-hypertensive drugs or nephrectomy to control their high blood pressure. The distinction between these two groups has been previously attempted by measuring levels of plasma angiotensin II, plasma renin or PRA. Such measurements, whilst complementary to saralasin infusion, do not appear to give as precise a division between the two groups. In addition, measurements of angiotensin II and PRA have the further disadvantage that the results are not immediately available.

The high blood pressure of patients who show no fall in blood pressure with saralasin before dialysis should be treated by gradual reduction in pre-dialysis weight. In most patients, this will reduce the blood pressure. If not, and particularly if blood pressure rises with weight removal, saralasin infusion should be repeated after dialysis. The arterial pressure of patients who show a fall in blood pressure with saralasin either before or after dialysis can be controlled by keeping the pre-dialysis weight steady and treating them with hypotensive drugs. In three such patients described here, the blood pressure was controlled by a combination of propranolol and hydrallazine. Bilateral nephrectomy was not required to control their blood pressure. This is consistent with the experience of our dialysis unit in the last 6 years. During this time, 140 patients have been treated by maintenance dialysis, and in none of the patients whose blood pressure was not controlled by weight reduction has it been necessary to perform a bilateral nephrectomy to control their blood pressure. If the pre-dialysis weight is kept constant and the weight gain between dialysis is restricted, their blood pressure can be controlled by hypotensive drugs. Eventually, the blood pressure becomes normal without hypotensive drugs. Presumably at this stage, they are no longer angiotensin II-dependent. Serial studies with saralasin in these patients will be of great interest.

To interpret the mechanism of the high blood pressure in those patients whose blood pressure fell with saralasin, it is important to remember that the vascular effect of angiotensin II is directly related to the fluid volume, or the salt intake [12, 13]. The three patients whose blood pressure fell with saralasin were selected because they appear to illustrate different facets of this relationship. Patient 3 had very high levels of angiotensin II and PRA predialysis. During a low infusion rate of saralasin, the lying blood pressure came down to normal, and upon standing there was a further fall in pressure. Post-dialysis, with a 1-kg reduction in weight, there was severe postural hypotension, without saralasin. This would suggest that pre-dialysis, the patient was already volume-depleted and that post-dialysis this became more pronounced. Patient 4, pre-dialysis, had a low normal value of PRA, with evidence of excess fluid. Predialysis, blood pressure was not angiotensin II-dependent. When 3 kg of fluid was removed during dialysis, there was a rise in blood pressure due presumably to the measured rise in PRA. Though the level of PRA achieved post-dialysis was not above the range achieved in sodium-depleted normals, the administration of saralasin at this time caused a fall in blood pressure, but not to normal. It would appear therefore that post-dialysis, the blood pressure became angiotensin II-dependent but, inasmuch as the blood pressure did not fall to normal levels with saralasin, it was also volume-dependent. The angiotensin II dependence was present, although the PRA was within the range of sodium-depleted normals, presumably because its vasoactivity was increased by persistent fluid excess. The mechanism causing the high blood pressure appears to have switched from volume dependency pre-dialysis, to a mixture of angiotensin II and volume dependency post-dialysis. In patient 5, with weight gain before dialysis, renin release was suppressed, giving values of PRA that were around the upper limit of normal. At this time, saralasin caused a small fall in blood pressure, but not to normal levels. This level of PRA, therefore, although lower than that usually associated with angiotensin II dependency, must have been inappropriately high for the presumed degree of volume excess. When further volume was removed during dialysis, there was a rise in PRA to a value above that achieved in sodium-deprived normals. A low infusion rate of saralasin then caused the blood pressure to fall to almost normal values. In this patient, the raised blood pressure pre-dialysis appears to have been maintained by a mixture of volume excess and angiotensin II, and post-dialysis by angiotensin II alone.

Summary

Five hypertensive haemodialysis patients have been infused with saralasin. The infusion appears to be a simple diagnostic test separating patients into two groups. First, there are those whose blood pressure does not fall with saralasin pre-dialysis, but does fall with weight removal during dialysis; the blood pressure in these patients can be controlled by a reduction in pre-dialysis weight. Second, there are those whose blood pressure does fall with saralasin either pre- or post-dialysis; their arterial pressure does not fall with weight removal, but can be controlled by anti-hypertensive drugs. In two of the patients who responded to saralasin, the mechanism of the high blood pressure appeared to change from volume dependency, partial or complete, with suppressed renin release, to angiotensin dependency, partial or complete, as weight was removed during dialysis. These patients illustrate the importance of the interaction between volume and the level of angiotensin II in the maintenance of hypertension.

Acknowledgements

We thank Mrs. N. MARKANDU, Mr. D. C. FAULKNER and Mr. J. C. KEDDIE for their help and Dr. R. E. KEENAN (Eaton Laboratories) for supplies of saralasin.

References

1 BROWN, J. J.; CURTIS, J. R.; LEVER, A. F.; ROBERTSON, J. I. S.; WARDENER, H. E. DE, and WING, A. J.: Plasma renin concentration and control of blood pressure in patients on maintenance dialysis. Nephron 6: 329–349 (1969).

2 VERTES, V.; CANGIANO, J. L.; BERMAN, L. B., and GOULD, A.: Hypertension in end-stage renal disease. New Engl. J. Med. *280:* 978–981 (1969).
3 SCHALEKAMP, M. A., et al.: Hypertension in chronic renal failure an abnormal relation between sodium and the renin-angiotensin system. Am. J. Med. *55:* 379–390 (1973).
4 ROSEN, S. M. and ROBINSON, P. J. A.: Interdependence of exchangeable sodium and plasma renin concentration in determining blood pressure in patients treated by maintenance dialysis. Br. med. J. *iv:* 139–143 (1973).
5 WEIDMAN, P. and MAXWELL, M. H.: The renin-angiotensin aldosterone system in terminal renal failure. Kidney int. *8:* S219–234 (1975).
6 BRUNNER, H. R.; GAVRAS, H.; LARAGH, J. H., and KEENAN, R.: Angiotensin II blockade in man by Sar[1]-Ala[8]-angiotensin II for understanding and treatment of high blood pressure. Lancet *ii:* 1045–1048 (1973).
7 STREETEN, D. H. P.; ANDERSON, G. H.; FREIBERG, J. M., and DALAKOS, T. G.: Use of an angiotensin II antagonist (saralasin) in the recognition of angiotensinogenic hypertension. New. Engl. J. Med. *292:* 657–662 (1975).
8 GEORGE, C. F.; LEWIS, P. J., and PETRIE, A.: Clinical experience with use of ultra sound sphygmomanometer. Br. Heart J. *37:* 804–807 (1975).
9 DUSTERDIECK, G. and MCELWEE, G.: Estimation of angiotensin II concentration in human plasma by radioimmunoassay. Some applications to physiological and clinical states. Eur. J. clin. Invest. *2:* 32–38 (1971).
10 ANDERSON, G. H.; FREIBERG, J. M.; DALAKOS, T. G., and STREETEN, D. H. P.: Agonistic response of vascular receptors to Sar[1]-Ala[8]-angiotensin II. Clin. Res. *23:* 218A (1975).
11 MACGREGOR, G. A. and DAWES, P. M.: Agonist and antagonist effects of Sar[1]-Ala[8]-angiotensin II in salt-loaded and salt-depleted normal man. Br. J. clin. Pharmac. *3:* 483–487 (1976).
12 BRUNNER, H. R.; CHANG, P.; WALLACH, R.; SEALEY, J. E., and LARAGH, J. H.: Angiotensin II vascular receptors: their avidity in relationship to sodium balance, the autonomic nervous system, and hypertension. J. clin. Invest. *51:* 58–67 (1972).
13 HOLLENBERG, N. K.; CHENITZ, W. R.; ADAMS, D. F., and WILLIAMS, G. H.: Reciprocal influence of salt intake on adrenal glomerulosa and renal vascular responses to angiotensin II in normal man. J. clin. Invest. *54:* 34–42 (1974).

Dr. G. A. MACGREGOR, Department of Medicine, Charing Cross Hospital Medical School, *London W6* (England)

Discussion[1]

POULSEN: As far as I can see, most investigators find an increase in the blood pressure when the plasma renin activity is low and a fall in blood pressure with high plasma renin concentrations when saralasin is injected. This, to me, suggests that the dose of saralasin is so high that its agonistic property is exhibited. Why do you not reduce the dose of saralasin so you have no measurable agonistic effect at low plasma renin activity? Wouldn't that make the interpretation much more simple?

MACGREGOR: As you may know, we have been trying to encourage the use of smaller amounts of saralasin. In fact, we have infused patients at much lower doses than that shown, e.g. starting at 25 ng/kg/min. We have never seen a fall in blood pressure at lower infusion rates, and in the majority of AII-dependent patients it requires an infusion rate of 1–2.5 µg/kg/min to lower blood pressure. In other words, lower doses or lower infusion rates of saralasin do not lower blood pressure.

BOYD: It seems to me that even with small doses where you get down to the bottom of the agonistic dose-response curve for saralasin, you still cannot be sure that the absence of response in any situation is not determined by a partial agonistic effect to saralasin, together with an antagonism of the existing effect of endogenous angiotensin II. If you cannot be sure of that, then an absent effect of saralasin in any state is extremely difficult to interpret.

ROBERTSON: Dr. MACGREGOR, have you ever experienced trouble, in patients with severe hypertension and chronic renal failure, where there is disproportionately high renin, in getting the renin levels down using propranolol, for example. There was a paper from Italy by MAGGIORE in 1975 in which he showed that, in this sort of situation, he was always able to get the renin down and the blood pressure under control with propranolol. Other workers have not always been able to do this and have, on occasions, had to resort to nephrectomy. Have you any comments on that point?

MACGREGOR: All AII-dependent patients were given oral propranolol to study

[1] Discussion of the papers of YAMAMOTO et al. and MACGREGOR and DAWES.

this. However, we found that long-term treatment with propranolol was not very effective in reducing blood pressure. There is a volume component as well in these patients. That is why we have chosen in clinical management of the patients to give propranolol and a vasodilator, namely hydrallazine.

MAXWELL: Dr. MACGREGOR, I just want to say that one must be very cautious in the interpretation of blood volume, extracellular water and so forth, in these patients who are severely anaemic, and have high cardiac outputs, inanition and protein deficiency. Until now, in the absence of saralasin tests, if a patient is on dialysis and hypertensive, I think the usual practice is to simply progressively salt deplete him and see if his blood pressure comes down. If it does not, one then gives antihypertensive drugs, and if it still does not, then one measures peripheral renin activity and if that is sky high, one considers bilateral nephrectomy. I do not see how the saralasin test changes this approach at all.

MACGREGOR: I think that there are two main benefits of saralasin infusion. Firstly, you can identify people who are definitely not AII-dependent at the time you infuse them and you can therefore say 'let's try more weight removal'. Secondly, in the AII-dependent patients, progressive volume depletion is unnecessary and makes them very ill. Some of the AII-dependent patients may need volume repletion as well as antihypertensive drugs to gain control of their blood pressure. Using this approach, bilateral nephrectomy has not been necessary.

SLATER: I remain a little unhappy about the specificity of this test, Dr. MACGREGOR. I was wondering whether you see a hypotensive response after total nephrectomy, sometimes?

MACGREGOR: The first slide showed an agonist response in an anephric patient.

SLATER: If you sodium-depleted an anephric patient, would you then get a response?

MACGREGOR: We did sodium-deplete the patient in the sense that we removed 2 kg of weight or volume, and the agonist response was still present. Of course, there had been no change in plasma renin activity.

FLETCHER: Could Dr. MACGREGOR give us the heart rate changes in his dialysis patients infused with saralasin?

MACGREGOR: During the saralasin infusion, there was no change in heart rate except when severe postural hypotension developed. Then there was an increase in standing pulse rate.

MIMRAN: I was very interested in the results of Dr. YAMAMOTO in three patients with phaeochromocytoma. We have also infused three people with phaeochromocytomas with saralasin, and one of them experienced an 11-mm Hg increase in blood pressure. The other two people did not change their blood pressure. I wonder if his findings are not a reflection of the effect of the 1-sarcosine-8-isoleucine analogue on the adrenal system. The adrenal agonistic effect of this compound may be greater than that of saralasin.

YAMAMOTO: This antagonist is also known to stimulate catecholamine secretion from the adrenal of dogs. However, as far as we checked, in 5 normal subjects, the urinary excretion of free catecholamines was reduced and not increased. However, neoplastic chromaffin tissue might be different from normal adrenal tissue.

FERNANDEZ-CRUX: I think that plasma renin activity in phaeochromocytoma can be normal. Can you tell us about your renin data?

Discussion

YAMAMOTO: We have three phaeochromocytoma patients who showed elevation of blood pressure. These subjects had a plasma renin activity higher than normal. We dealt with two other subjects with phaeochromocytomas during the normotensive phase. These two subjects also showed a somewhat elevated plasma renin activity, but yet they did not show elevation of blood pressure with this compound. Dr. YOSHINAGA has further experience in this area.

YOSHINAGA: In our experience, about half of the patients with phaeochromocytoma had a high plasma renin activity.

Angiotensin Antagonists as Diagnostic and Pharmacologic Tools[1]

WILLIAM A. PETTINGER[2] and HELEN C. MITCHELL

The University of Texas Southwestern Medical School, Departments of Pharmacology and Internal Medicine, Division of Clinical Pharmacology, Dallas, Tex.

Contents

Introduction	203
Pharmacokinetics of Saralasin	205
Diagnostic Use of Saralasin	206
Hypotensive Response	208
Augmentation of Split Renal Vein Renins	209
Autonomous Aldosterone Secretion	210
Angiotensin Antagonists as Tools in Clinical Pharmacology	211
Summary	212
References	212

Introduction

Angiotensin antagonists could be useful diagnostically, therapeutically, or in the study of disease and drug mechanisms purportedly mediated by the renin-angiotensin axis. The task of describing the state of the art in this area would be rendered considerably easier if the pathogenetic role of the renin-angiotensin axis were more clearly understood. Some of the physiological roles of angiotensin have been identified, such as angiotensin vasoconstriction and angiotensin mediation of aldosterone secretion. However, various factors can alter the response to angiotensin and can

[1] Support was provided by The Texas Heart Association, NIH Grant No. 06-D-00007, and the Norwich Pharmacal Company.
[2] Dr. Pettinger is a Burroughs Wellcome Scholar in Clinical Pharmacology.

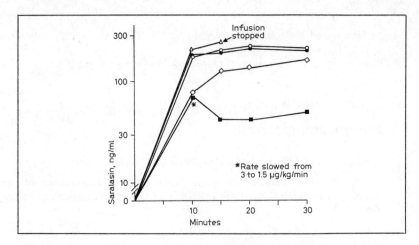

Fig. 1. Plasma saralasin concentration as a function of infusion-duration and rate. Infusion rate of the top four curves (4 patients) was 10 μg/kg/min and in the bottom (1 patient), it was 3 μg/kg/min initially. Saralasin plasma concentrations were measured by radioimmunoassay as previously described [1].

replace angiotensin as the dominant mediator of vasoconstriction or mineralocorticoid stimulation. For example, when the source of renin is removed by nephrectomy, blood pressure can be controlled normally and plasma aldosterone is maintained at nearly normal levels.

In the study of hypertension, the logic is further complicated by the multitude of mechanisms which are involved in blood pressure maintenance. We can simplify them or oversimplify them using vague terms such as 'volume' or 'vasoconstrictor' mechanisms, which are useful but imprecise. The well-known neuroendocrine mechanisms maintaining blood pressure are the adrenergic nervous system, the renin-angiotensin axis and antidiuretic hormone. These control mechanisms participate in blood pressure maintenance through a variety of channels, some of which are direct and others very indirect. Additionally, other factors such as sodium balance and distribution participate in very fundamental ways in maintaining blood pressure at high, normal or low levels. Induction of a net negative sodium balance results in lowering of blood pressure in almost every type of hypertension. A proponent of the sodium thesis for hypertensive mechanisms might thus argue that sodium is a cause of all hypertension. Additionally, α-adrenergic blockade with drugs such as phentolamine or phenoxybenzamine results in blood pressure lowering in the majority of patients, particularly when the drugs are administered intravenously. Thus, there are some who might argue that abnormalities of the sympathetic nervous system are the major mediators of high blood pressure. More recently, certain patients have been demonstrated to respond with a fall in blood pressure when angiotensin antagonists are given. Thus, proponents of angiotensin mechanisms for high blood pressure have voiced their opinions as well. When antidiuretic hormone antagonists become available for clinical study, we should anticipate vasopressin theories of hypertension coming to the fore. When blood pressure falls as a result of induction of a net sodium loss or administration of a neuroendocrine blocker, one might intimate that this particular mechanism is maintaining blood pressure. However, such a result does not establish a particular mechanism as the sole or primary mediator of one type of hypertension. With this perspective of blood pressure maintaining mechanisms we will discuss the scientific rationale for potential use of angiotensin antagonists.

Pharmacokinetics of Saralasin

Before embarking on discussins of mechanisms, let us review some pharmacokinetics of the drug saralasin, since this information is a prerequisite to intelligent experimental design. In man, the half-life of saralasin in plasma is 3.1 min, while its effective half-life (with respect to blood pressure effects) is 4–8 min [1]. Therefore, the duration of a constant infusion required to achieve a plateau of plasma concentration is 10–15 min (i.e. 4–5 ×half-life) as shown in figure 1. The average half-life required for return of blood pressure after low bolus doses was 4–5 min [1]; after discontinuing infusions it was 7.8–8.5 min [1]. More recent results using higher bolus doses are consistent with the longer half-life following infusions [2]. Some individual patient variability occurs in plasma saralasin concentrations at equivalent infusion rates (fig. 2). However, far greater variability exists in the individual's response to a given plasma saralasin concentration (fig. 3). Some patients had marked hypotensive responses to plasma concentrations of 3 ng/ml while others required more than 100 ng/ml to produce a similar degree of hypotension. Because of this degree of variability in response, we have shifted from bolus injections to infusions when doing more sophisticated clinical pharmacologic studies [1, 3].

There are qualitative as well as quantitative variabilities in blood pressure response. Blood pressure tends to go up initially after a bolus

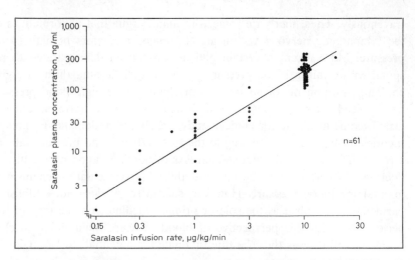

Fig. 2. Saralasin plasma (venous) concentration after 30-min infusions in 12 subjects at rates ranging from 0.15 to 30 μg/kg/min. Infusion rates were gauged by drip counting using a 'Volutrol' system which may account for some of the variability in plasma concentration.

injection or when starting an infusion. The initial elevation represents an 'intrinsic' action of saralasin. It is greater in non-responders [1, 2] and with higher doses by bolus injection and may be dose-related. We blocked this 'intrinsic' action of higher saralasin doses in animals by giving a small dose 10 min prior to the large bolus [4]. The same technique may be useful in preventing elevations of blood pressure of up to 230/130 mm Hg which can occur with high-dose bolus injections in hypertensive man [2].

The pharmacokinetics relevant to the effects of saralasin on other angiotensin receptors are different. An infusion time of more than 30 min and less than 60 min is required to maximally suppress angiotensin-mediated aldosterone secretion in man [5]. Blockade of the angiotensin-mediated short-loop (fig. 4) inhibitory to renin release has a rapid onset [1, 7] but persists for more than 1 h (fig. 5). This effect of saralasin on the short-loop is blocked by propranolol in hypertensive man [3] and in rats [6].

Diagnostic Use of Saralasin

A major thrust of clinical investigators using angiotensin antagonists is in the direction of its use as a screening test for patients with renal ar-

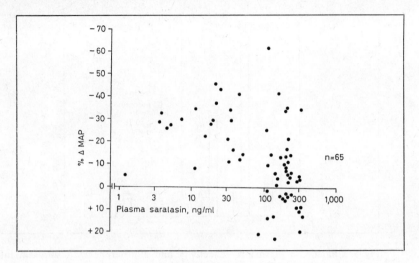

Fig. 3. Relationship between saralasin plasma concentration and maximum change in blood pressure with saralasin infusions.

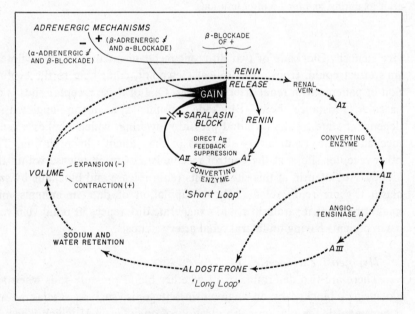

Fig. 4. Diagram of renin release control mechanisms. See text for discussion. Angiotensin III [4], α-adrenergic inhibitory receptors [9, 12, 13], and short-loop mechanism [6, 14, 15] are discussed elsewhere.

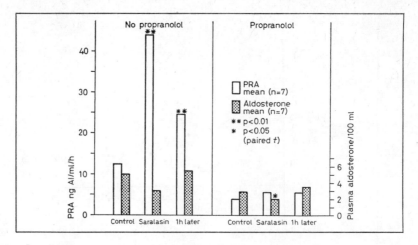

Fig. 5. Changes in plasma renin activity and aldosterone with saralasin infusion (30 min) without and with propranolol in seven vasodilatory drug-treated subjects. Aldosterone suppression with 30-min infusion is distinctly less than after 1-hour infusions [5]. Note the lowering and stabilization of both the renin and aldosterone with propranolol administration even though subjects are supine 30 min prior to, during, and for 1 h after infusions.

tery stenosis. Blockade of two different mechanisms or sites of angiotensin's effect could, from a theoretical viewpoint, contribute to the evaluation of patients with renal artery stenosis. One of these receptor sites is on vascular resistance beds. Blockade at this site in an angiotensin-dependent state results in blood pressure lowering, which implies a renin mechanism (i.e. renal artery stenosis) in the patient's hypertension. The other receptor site is at the control mechanism of renin release within the kidney. Angiotensin at this site inhibits renin release and blockade by saralasin triggers renin release. By extrapolation of data in animals, one might predict that saralasin could exaggerate differences in renal vein renins in patients having unilateral renal artery stenosis.

Hypotensive Response

There are two determinants of whether blood pressure falls when angiotensin is blocked by an angiotensin antagonist such as saralasin. One determinant is the plasma concentration of angiotensin II which is generally related to serum renin activity. The higher the serum renin activity, the greater the tendency for blood pressure to fall when angiotensin vasoconstriction is blocked. The second determinant of whether blood pressure falls when angiotensin is blocked is the state of fluid and electrolyte balance which determines the 'effective blood volume'. The 'effective blood volume' is a poorly defined quantity. It is related to the volume of blood between the aortic valve and the proximal or afferent areriloes in the peripheral circulation. Expansion of this volume occurs in over-hydrated states, in circumstances of excess salt intake, and with most antihypertensive drugs when not accompanied by diuretic agents. The expansion of this volume tends to nullify the blood pressure lowering effect of nearly all blood pressure lowering interventions including the angiotensin antagonist saralasin. In fact, patients with extremely high plasma renin activities which have over-expansion of this volume space may have no response to saralasin at all [1]. Also, patients having surgically correctable renal artery stenosis with high renin activity may also have no hypotensive response to saralasin [7]. Alternatively, contraction of this space by diuretic agents results in enhanced blood pressure lowering by angiotensin antagonists [1, 8]. These two major determinants, i.e. plasma renin activity and 'effective blood volume', thus determine whether blood pressure falls when angiotensin antagonists are administered. A thorough and systematic investigation of these two determinants should be done in various disease states in which angiotensin antagonists might be used for diagnostic purposes.

Augmentation of Split Renal Vein Renins

The rate of renin release is determined by a number of factors or mechanisms which are illustrated schematically in figure 4. A primary and major determinant of the rate of renin release is a volume-sensitive intrarenal mechanism illustrated by the long loop in this figure. That is, angiotensin-mediated aldosterone secretion induces sodium and water retention and expansion of the extracellular fluid volume in which the volume-sensitive mechanism participates. Expansion of this volume results in suppression of renin release or, alternatively, contraction of this volume – as in salt-depleted states or when diuretics are administered – results in stimulation of renin release. Stimulation or suppression of renin release by this mechanism is influenced to some degree by the adrenergic nervous system, particularly in the upright position. However, in the supine position, the adrenergic nervous system has relatively little effect. Therefore, the administration of propranolol suppresses renin release considerably less in patients in the supine than in the standing position.

We have done extensive studies in animal models in which various components of these control mechanisms have been standardized and characterized. β-Adrenergic receptors mediate stimulation of renin release and alpha adrenergic receptors suppress renin release [9–13]. Angiotensin *per se* has a direct effect, indicated in figure 4 by the short loop, in suppressing renin release [6, 14, 15]. When these various control mechanisms are combined in any animal model, they result in much more than additive or synergistic effects (described here as *gain*). Now, in the patient with arterial stenosis of one kidney, its signal is taken to indicate that the volume is contracted, thus resulting in a greater renin release than would occur ordinarily for a given volume state. Because of this, the renin release tends to be higher from the side of the lesion. However, if the renal artery stenosis is moderate (40–50%) and the volume is sufficiently expanded so that there is no pressure gradient across the renal artery stenosis at the time the study is done, the rate of renin release from the side of the lesion may be equivalent to that on the opposite side. Previously, hypotensive interventions have resulted in vasodilation of both renal and peripheral vascular beds. The vasodilation increases flow across the stenotic lesion as well as on the contralateral side, and activates the β-adrenergic limb of the *gain*. Renin release is thereby enhanced from both kidneys, but the increase tends to be greater from the kidney on the side of the lesion. From the theoretical considerations illustrated in figure 4, saralasin infusion could possibly achieve a similar effect without the magnitude of hypotension occurring with vasodilating drugs.

Autonomous Aldosterone Secretion

Angiotensin antagonists have the capacity of blocking angiotensin-mediated aldosterone secretion in man (fig. 5) [5, 16]. An infusion time of more than 30 min is required to suppress plasma aldosterone with saralasin in patients, a near maximum effect being achieved by 60 min. We have studied a series of high and normal renin patients with saralasin and found that aldosterone in the plasma was as high in the normal renin group as in the high renin group [5]. Surprisingly, however, saralasin inhibited aldosterone secretion to an equivalent extent in both groups. Therefore, some mechanism is preventing aldosterone from rising in high renin patients; yet, the rate of aldosterone secretion from the adrenal cortex is still angiotensin-dependent. Alternatively, one could argue that states of primary aldosterone excess may be independent of angiotensin mediation of aldosterone secretion, almost by definition. Perhaps normal aldosterone secretion is sustained by unusually low levels of angiotensin in 'low renin' hypertension. This question could be answered by the use of angiotensin antagonists. Whether saralasin blockade of angiotensin-mediated aldosterone secretion would result in suppression of plasma aldosterone in patients with primary aldosteronism has not yet been reported. The response to saralasin infusion could conceivably distinguish patients with adrenal adenomas from those with hyperplasia, or possibly patients with low renin hypertension from those with primary aldosteronism.

Angiotensin Antagonists as Tools in Clinical Pharmacology

The use of angiotensin antagonists as tools for understanding drug mechanisms is of particular interest to us [3]. Saralasin infusion could induce some lowering of blood pressure to hypotensive levels in subjects treated with vasodilator drugs [3, 17]. Thus, maintenance of blood pressure was angiotensin-dependent in those patients. Vasodilator drugs induce renin release, which can be impaired by propranolol. When propranolol was added to the vasodilator drugs, blood pressure and plasma renin activity were lower and the hypotensive response to saralasin was reduced or ablated [3]. Thus, propranolol administration was associated with reduction of renin release and a shift of the blood pressure maintaining mechanism from angiotensin to other mechanisms. Additionally, there was a precise correlation ($p<0.001$) between plasma renin activity and the magnitude of saralasin-induced blood pressure lowering. Thus, reduction of renin release by propranolol is directly related to reduction of angiotensin maintenance of blood pressure at elevated levels; this is a major mechanism whereby propranolol contributes to the hypotensive action of the vasodilator-β-blocker drug interaction in hypertensive man.

In conclusion, angiotensin antagonists have a demonstrated potential in the study of the roles of the renin-angiotensin system in pharmacology and physiology. Whether these agents will have sufficient precision for clinical use in screening for, and evaluation of, patients having renal artery stenosis is yet to be determined. Some very important quantities concerning the incidence of false positive and, particularly, false negative results are absent from current reports [2, 18] and are yet to be obtained.

Further systematic studies are required to determine the optimal diuretic stimulus in a broad-based population to quantify the potential for erroneous conclusions.

Summary

Angiotensin antagonists have become useful tools in studying pharmacologic and pathophysiologic roles of the renin-angiotensin axis. Several of these uses are described herein. Their value as tools in the diagnosis of renal artery stenosis is yet to be determined.

References

1 PETTINGER, W. A.; KEETON, K., and TANAKA, K.: The radioimmunoassay and pharmacokinetics of saralasin (1-Sar-8-Ala-angiotensin II) in the rat and hypertensive man. Clin. Pharmac. Ther. *17:* 146–158 (1975).
2 MARKS, L. S.; MAXWELL, M. H., and KAUFMAN, J. J.: Saralasin bolus test. Rapid screening procedure for renin-mediated hypertension. Lancet *ii:* 784–787 (1975).
3 PETTINGER, W. A. and MITCHELL, H. C.: Renin release, saralasin and the vasodilator-beta blocker drug interaction in man. New Engl. J. Med. *292:* 1214–1217 (1975).
4 CAMPBELL, W. B.; BROOKS, S. N., and PETTINGER, W. A.: Angiotensin II- and angiotensin III-induced aldosterone release *in vivo* in the rat. Science *184:* 994–996 (1974).
5 PETTINGER, W. A. and MITCHELL, H. C.: Clinical pharmacology of angiotensin antagonists. Fed. Proc. Fed. Am. Socs exp. Biol. (in press, 1976).
6 KEETON, T. K.; PETTINGER, W. A., and CAMPBELL, W. B.: The effects of altered sodium balance and adrenergic blockade on renin release induced by angiotensin antagonism. Circulation Res. *38:* 531–539 (1976).
7 DONKER, A. J. M. and LEENEN, F. H. H.: Infusion of angiotensin II analogue in two patients with unilateral renovascular hypertension. Lancet *ii:* 1535–1537 (1974).
8 GAVRAS, H.; BRUNNER, H. R.; LARAGH, J. H.; SEALEY, J. E.; GAVRAS, I., and VUKOVICH, R. A.: An angiotensin converting-enzyme inhibitor to identify and treat vasoconstrictor and volume factors in hypertensive patients. New Engl. J. Med. *291:* 817–821 (1974).
9 PETTINGER, W. A.; AUGUSTO, L., and LEON, A. S.: Alteration of renin release by stress and adrenergic receptor and related drugs in unanesthetized rats; in BLOOR Comparative pathophysiology of circulatory disturbances, pp. 105–117 (Plenum Publishing, New York 1972).
10 PETTINGER, W. A.; CAMPBELL, W. B., and KEETON, K.: Adrenergic component

of renin release induced by vasodilating antihypertensive drugs in the rat. Circulation Res. *33:* 82–86 (1973).
11 TANAKA, K. and PETTINGER, W. A.: Renin release and ketamine-induced cardiovascular stimulation in the rat. J. Pharmac. exp. Ther. *188:* 229–233 (1974).
12 PETTINGER, W. A.; CAMPBELL, W. B.; KEETON, T. K., and HARPER, D. C.: Evidence for a renal alpha adrenergic receptor inhibiting renin release. Circulation Res. *38:* 338–346 (1976).
13 VANDONGEN, R. and PEART, W.: The inhibition of renin secretion by alpha-adrenergic stimulation in the isolated rat kidney. Clin. Sci. mol. Med. *47:* 471–479 (1974).
14 VANDER, A. J. and GEELHOED, G. W.: Inhibition of renin secretion by angiotensin II. Proc. Soc. exp. Biol. Med. *120:* 399–403 (1965).
15 PETTINGER, W. A. and KEETON, K.: New receptor mechanisms controlling renin release; in ONESTI *et al.* Regulation of blood pressure by central nervous system (Grune & Stratton, New York 1976).
16 GAVRAS, H.; BRUNNER, H. R.; LARAGH, J. H.; GAVRAS, I., and VUKOVICH, R. A.: The use of angiotensin-converting enzyme inhibitor in the diagnosis and treatment of hypertension. Clin. Sci. mol. Med. *48:* 001s–004s (1975).
17 PETTINGER, W. A. and KEETON, K.: Hypotension during angiotensin blockade with saralasin. Lancet *i:* 1387–1388 (1975).
18 STREETEN, D. H. P.; ANDERSON, G. H.; FREIBERG, J. M., and DALAKOS, T. G.: Use of an angiotensin II antagonist (saralasin) in the recognition of 'angiotensinogenic' hypertension. New Engl. J. Med. *292:* 657–662 (1975).

WILLIAM A. PETTINGER, MD, Division of Clinical Pharmacology, The University of Texas Health Science Center, 5323 Harry Hines Blvd., *Dallas, TX 75235* (USA)

The Use of Saralasin in the Recognition of Angiotensinogenic Hypertension[1]

David H. P. Streeten, Theodore G. Dalakos and Gunnar H. Anderson, jr.

Section of Endocrinology, Department of Medicine, State University of New York Upstate Medical Center, Syracuse, N.Y.

Contents

Introduction	214
Method of Use of Saralasin	216
Need for Antecedent Na Loss	216
Method of Administration of Saralasin	218
Significance of Hypotensive Response to Saralasin	219
Nature of Renal and Renovascular Lesions in Saralasin Responders	222
Summary	224
References	224

Introduction

Evidence that any given biochemical abnormality is the cause of hypertension becomes convincing if: (1) the abnormality can be shown to be present consistently when the blood pressure is elevated; (2) specific antagonism of the abnormal substance lowers the blood pressure to normal; (3) removal or prevention of production of the abnormal substance corrects

---

1   Supported by a Research Grant (HL14076) from the National Institute of Heart and Lung Disease, a Graduate Training Grant in Endocrinology (AM07146) from the National Institute of Arthritis, Metabolism and Digestive Diseases, A Clinical Research Center Grant (RR00229) from the Division of Research Facilities and Resources, USPHS, and grants from the Central New York Regional Medical Program and Eaton Laboratories, Norwich, N.Y.

the hypertension, and (4) administration of the incriminated compound can be shown to elevate the blood pressure once again. By satisfying such modified Koch's postulates, several distinct compounds, when produced in excessive amounts, have been shown to be the cause of hypertension in man. There can be little doubt, for instance, that the hypertension seen in patients with phaeochromocytomas results from excessive catecholamine activity since (1) elevated levels of epinephrine and/or norepinephrine can be demonstrated in urine and plasma, (2) antagonism of the action of these catecholamines by such specific agents as phenoxybenzamine and phentolamine lowers the blood pressure to normal, (3) surgical removal of the tumour(s) reproducibly cures the hypertension, and (4) intravenous infusions of epi- and norepinephrine will invariably raise the blood pressure to at least its previous levels. The role of excessive blood levels of aldosterone in the hypertension of primary aldosteronism satisfies the same criteria for a cause and effect relationship: hyperaldosteronism is demonstrable by measurement, its antagonism by spironolactone and its correction by surgical adrenalectomy almost invariably overcome the hypertension and, though this is less readily shown in human patients, the administration of excessive amounts of mineralocorticoids will raise the blood pressure again.

In recent years, there has been increasing evidence of the same type that excessive angiotensin II (AII) activity is the cause of hypertension in some patients. Thus, plasma levels of AII or of renin activity (PRA) are elevated in some patients either with renal arterial stenosis or with renal parenchymal disorders associated with hypertension [1, 2]. Antagonism of AII by the highly specific antagonist, saralasin, in many individuals of this type will lower the blood pressure to normal [3–5], and correction of the hyperangiotensinaemia by surgical correction of the renal arterial stenosis or by unilateral nephrectomy will 'cure' the hypertension in such patients [6–8]. Intravenous infusions of authentic angiotensin II will invariably raise the blood pressure to an extent proportional to the rate of infusion of the peptide [9, 10]. Thus, patients whose hypertension can be shown by these criteria to have convincing cause-and-effect relationships with excessive levels of AII or PRA, have been considered to have 'angiotensinogenic' hypertension. This designation distinguishes the above group of patients from individuals who may be hypertensive and purely coincidentally hyperangiotensinaemic. Examples of this phenomenon are those hypertensive females who happen to have elevated plasma renin and angiotensin levels because they are receiving oestrogen-progestogen combinations, yet whose blood pressure is not raised by the hyperangiotensinaemia since it does not return to normal either when saralasin is infused or when cessation of the intake of oral contraceptives restores plasma renin activity to normal.

The existence of angiotensinogenic hypertension has been suspected for several years, in patients with unilateral renal arterial stenosis associated with elevations of PRA in the ipsilateral renal vein and subsequent restoration of the blood pressure to normal by surgical correction of the renal arterial stenosis. Several groups of investigators have shown that surgical cure of the hypertension could be reliably predicted in a very high percentage of such patients provided that the renal vein PRA ratio exceeded 1.5 [7, 8, 11]. Yet it was possible that surgical correction of the renal arterial stenosis might have restored the blood pressure to normal in those patients not by lowering PRA – which it certainly did – but by some other mechanism, such as increasing prostaglandin production or potentiating sodium excretion. The great potential value of a specific AII antagonist in these circumstances was that it might provide additional confirmatory evidence that the blood pressure would decrease after surgery, the level falling to approximately the same extent as it fell during preoperative administration of the AII antagonist. The extent to which saralasin has been helpful in providing rapid and reliable evidence of the presence of an 'angiotensinogenic' component in the hypertension of individual patients will be examined in this paper.

Method of Use of Saralasin

Need for Antecedent Na Loss

Studies in rats with 'Goldblatt hypertension' have shown that there is an initial phase of renin-dependent hypertension which is of very short duration in the 'one-kidney' model (i.e. unilateral renal arterial constriction and contralateral nephrectomy) and lasts at least five weeks in the 'two-kidney' model [12–15]. In each type of Goldblatt hypertension, the renin dependency seemed to be overcome or obscured by progressive retention of sodium and water, the correction of which restored obvious renin dependency as manifested by a good fall in blood pressure, often to the normal range, during saralasin infusion. We have confirmed the relevance of these findings to human patients with hypertension. Figure 1 shows blood pressure responses to saralasin in a patient with known, uni-

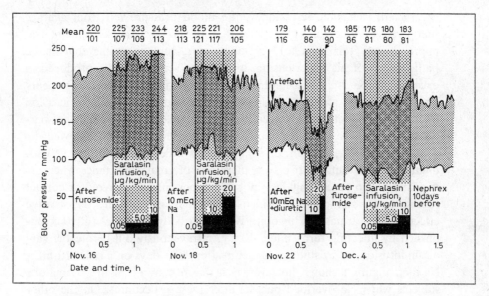

Fig. 1. Effects of saralasin infusions on the blood pressure in a hypertensive male, aged 42, with renal disease associated with excessive renin production from the left kidney, and severe peripheral oedema. Saralasin induced a rise in blood pressure after furosemide injection on November 16, but lowered the blood pressure slightly (from 218/113 to 206/105) on November 18 after restriction of Na intake to 10 mEq daily for three days. A dramatic fall in blood pressure (from 179/116 to 140/86) occurred after Na restriction and chlorthalidone administration for 4 days, on November 22. After left nephrectomy, blood pressure fell to 185/86 and was unaffected by saralasin (on December 4). Oedema was present on November 16 and 18, absent on November 22 and December 4. Recording of systolic blood pressure on November 22 was artefactually lowered by an erroneous setting of the Arteriosonde, which did not affect the diastolic pressures.

lateral, renal disease associated with severe and significant ipsilateral elevation of renal vein PRA (ratio abnormal/normal = 1.85), and severe oedema. He experienced no fall (in fact, there was a rise) in blood pressure during i.v. infusion of saralasin preceded by an intravenous injection of furosemide (40 mg). After therapy with a 10-mEq Na diet alone for three days, saralasin produced a slight, somewhat equivocal fall in blood pressure, but after the diet together with a diuretic (chlorthalidone, 100 mg daily) for four days, the blood pressure became responsive to saralasin, falling from 179/116 to 140/86 and 142/90 mm Hg during the infusion of saralasin at 10 and 20 µg/kg/min. After excision of the small, contracted, ischaemic left kidney, the blood pressure fell from average preoperative values of 215/112 to 185/86 mm Hg and has remained at approximately this level for several months, confirming the role in the pathogenesis of his hypertension, of renin or something else produced in excess by the ischaemic kidney. Similarly, in hypertensive patients who have shown an excellent hypotensive response to saralasin infusion after pre-treatment with intravenous furosemide, we have found that partial or complete loss of the hypotensive response to saralasin followed the administration of 2,000 ml, or sometimes as little as 500 ml of 0.9% sodium chloride solution [16]. It is, therefore, of great importance when using saralasin to detect angiotensinogenic hypertension, to avoid or correct sodium and water overload which may, otherwise, obscure hypotensive effects of the saralasin. This objective may usually be successfully accomplished by injecting furosemide, 40 mg, intravenously 3 h before the saralasin infusion, or by studying patients after three days on a strict 10-mEq Na diet. Figure 1 shows, however, that failure may occur in oedematous subjects when Na overload persists after these procedures. On the other hand, the avoidance of excessive natriuresis may also be important, since it is theoretically possible, in hypertension originally of central origin, that profound natriuresis might convert the mechanism of the hypertension into an angiotensinogenic one, with consequently misleading diagnostic implications. However, this latter theoretical possibility is certainly of far less consequence in the diagnostic use of saralasin than the possibility of overlooking a surgically remediable renal lesion by failure to accomplish adequate correction of pre-existing sodium overload before the saralasin test.

Method of Administration of Saralasin

When given by intravenous injection or infusion, saralasin may lower, raise or fail to change the blood pressure. Administration of the drug during or shortly after therapy with other potent antihypertensive medications, will often cause the blood pressure to fall to dangerously low levels [17] and for this reason all antihypertensive medications (apart from diuretics) should be withheld for several days (preferably 1–3 weeks) before saralasin is given. When saralasin has been infused after furosemide only (40 mg intravenously) we have seen the blood pressure fall *below* normal limits in only 1 of over 900 hypertensive patients studied in this way. Thus, in the absence of other drug therapy, the danger of precipitating hypotension by saralasin administration is remote. On the other hand, saralasin raises blood pressure in many hypertensive subjects. The rise may be transient, lasting less than 3 min, as first reported by WALLACE et al. [18], or it may continue as long as the saralasin is infused, showing dose-response relationships and having the characteristics of an AII-like agonistic response, as ANDERSON et al. [19] have described. The latter agonistic actions of saralasin are usually mild, never endure for more than a few minutes after discontinuing the saralasin infusions, and have never been observed to have any immediate or delayed adverse effects, in our experience. It is probably easier to avoid any harmful effects from such hypertensive actions of saralasin, if the drug is first infused at a low rate (0.05–0.2 µg/kg/min) since the pressor effects are milder at low infusion rates. If the blood pressure rises during infusions of saralasin at these low rates, the infusion can readily be stopped to avoid potentially more serious elevations in pressure from higher rates of infusion. Transient rises in blood pressure occur frequently, too, after intravenous *injections* of saralasin, when somewhat longer agonistic responses are quite commonly observed. The transient responses are readily evident in the report of MARKS et al. [20] on the diagnostic use of saralasin 'bolus' injections. The more prolonged agonistic responses which we encountered [19] after saralasin injections led us to suggest that intravenous injections of saralasin are too dangerous to warrant their use [21], a view shared by MACGREGOR and DAWES [22]. However, the effects of injections of saralasin are frequently clear-cut, dramatic, and potentially useful, if some means of avoiding adverse consequences of the frequent pressor effects can be found. In general, the safest and most satisfactory way to administer saralasin appears to be after preliminary natriuresis (e.g. 3 h after furosemide, 40 mg intravenously), starting with a slow rate of infusion (0.05–0.2 µg/kg/min) for 10–15 min and increasing thereafter, unless pressor effects are observed, to 5 or 10 µg/kg/min for at least another 15–20 min while recording blood pressure every 1 or 2 min. These tests should be preceded by a control period of at least 15–30 min and should be conducted while the patient is recumbent, in a quiet environment, devoid of distracting influences.

Significance of Hypotensive Response to Saralasin

When we infused saralasin, as described above, into 26 normal subjects, 3 h after intravenous furosemide (40 mg), the blood pressure changed little or not at all in most individuals. No normal subject experienced a fall in blood pressure greater than 10/8 mm Hg.

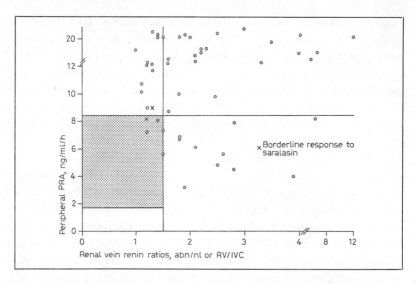

Fig. 2. Peripheral PRA levels and renal vein PRA ratios (abnormal/normal side or renal vein/inferior vena cava) in 50 hypertensives whose blood pressure fell by >10/8 mm Hg during saralasin infusion ('responders'). The predominance of elevated peripheral PRA values and of renal vein PRA ratios 1.5 or greater, is evident.

Among 600 unselected hypertensive patients given saralasin as described, at least three weeks after stopping their antihypertensive medications, 62 (10.3%) showed a fall in blood pressure of more than 10/8 mm Hg. Whenever such an excessive hypotensive effect of saralasin was observed, we have attempted to measure peripheral PRA (at 11 a.m., after furosemide, 40 mg intravenously, at 8 a.m. and after standing and walking slowly from 9 to 11 a.m.), and renal vein PRA levels, usually at 2 p.m. the next day after another 8-a.m. injection of furosemide and after tilting the patient, head up, to 30–45 degrees for 15 min. The results of such peripheral and renal vein PRA measurements in 50 of the 62 patients who showed a hypotensive response to saralasin, are shown in figure 2. It is evident that: (a) the renal vein PRA ratio (either abnormal side/normal side or higher side/inferior vena cava) was 1.5 or higher in all except 15 of the patients; (b) peripheral PRA was elevated in 12 of these 15 patients, and (c) in only 3 patients were the renal vein PRA ratios below 1.5 and the peripheral vein PRA levels normal.

Thus, 47 of 50 (94%) hypotensive responders to saralasin had renal vein PRA ratios above 1.5 and/or elevated peripheral levels of PRA. These findings confirm previous observations on our first 221 patients

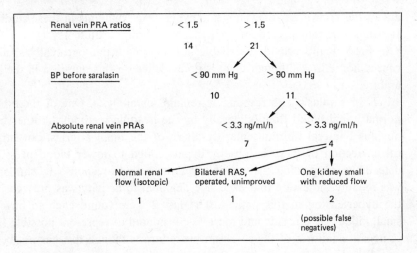

Fig. 3. Plasma renin activity results in saralasin non-responders (n = 35).

[5], and indicate that when there is a hypotensive response to saralasin, there is a very strong possibility that peripheral and/or renal vein PRA measurements will be abnormal. These findings, therefore, provide strong evidence that a hypotensive response to saralasin indicates that excessive angiotensin II activity might well be responsible for part or all of the elevation in blood pressure.

It is extremely difficult to be certain whether some of the patients who failed to experience a fall in blood pressure during saralasin might actually have had angiotensinogenic hypertension which was not recognized by the saralasin procedure. In an attempt to study this possibility, the results of renal vein PRA ratios and other studies in 35 patients who did *not* show a hypotensive response to saralasin infusion have been summarized in figure 3.

Renal vein PRA ratios were below 1.5 in 14 patients including 4 of the 5 whose perpheral PRA was elevated. Among the 21 patients whose renal vein PRA ratios were 1.5 or more, it was found that diastolic blood pressure had fallen to 90 mm Hg or less, in consequence of the preceding furosemide injection and before the saralasin infusion, in 10 patients. Since saralasin almost never lowers blood pressure below normal levels, these failures to respond to saralasin cannot, therefore, be interpreted. Inspection of the observations in the remaining non-responders showed that the absolute level of renal vein PRA in 7 patients was below 3.3 ng/ml/h – which was the lowest absolute level of renal vein PRA recorded in any of the known angiotensinogenic hypertensive patients in these studies. It seems reasonable to doubt whether such low absolute levels of renal vein PRA would provide a reliable indicator of a pathogenically significant abnormality, although there is no evidence on this question in the literature.

Four saralasin non-responders remain for analysis. One of these had normal renal blood flow bilaterally by the isotopic method. Another had obvious bilateral fibromuscular dysplasia of the renal arteries, but surgical correction of these lesions by bypass failed to lower blood pressure detectably, supporting the possibility that the non-response to saralasin was reliably indicating that excessive angiotensin activity was not causing the hypertension in this patient. The last 2 were found each to have a small kidney on one side and must be considered to represent possible false-negative failures of the saralasin procedure, although this has not been established conclusively since the hypertension in these patients has not been shown to be cured by surgery. Thus, there is a reasonable possibility that the saralasin procedure might have failed to indicate correctly the angiotensinogenic nature of the hypertension in only 2 of these non-responders.

These observations indicate that when blood pressure fails to fall during saralasin infusion, there is only a small chance that a truly angiotensinogenic type of hypertension is being overlooked. In view of the fact that sodium overload can undoubtedly obscure a hypotensive response to saralasin, it is possible that use of saralasin after more effective natriuresis than can be accomplished by a single intravenous injection of furosemide might further reduce the incidence of apparently false-negative responses.

Nature of Renal and Renovascular Lesions in Saralasin Responders

As far as possible, all 62 hypotensive responders among the first 600 patients whose blood pressure responses to saralasin were studied have been further evaluated with measurements of creatinine clearance, intravenous pyelograms, isotopic renal flow studies, renal vein PRA measurements, bilateral selective renal arteriography, and other investigations. The types of disorder revealed by these studies were as follows:

1) Unilateral renal arterial stenosis with 'adequate' creatinine clearance (>50 ml/min), and normal or borderline serum creatinine concen-

tration: 18 patients (29%). 13 of these patients, or 21% of the entire group, have been subjected to surgery (renal arterial bypass, unilateral nephrectomy or heminephrectomy), and diastolic blood pressure is below 92 mm Hg with no treatment in all of the operated patients 2–24 months after surgery, except in one patient who was heminephrectomized because the PRA was high only in the vein from the lower segment of one kidney. Renal arterial bypass surgery was unsuccessful in 3 of these patients who then required unilateral nephrectomy which was in all instances followed by a fall of blood pressure to normal. One of the 5 remaining individuals in this group of 18 patients died a few days before renal arterial surgery could be performed, and surgery has been declined or considered undesirable in 4 patients.

2) Bilateral renal disease (parenchymal lesions or renal arterial constrictions) with azotaemia was present in 13 patients (21% of the group). Most of these patients have been changed from their previous forms of medical therapy to propranolol with or without other medications (usually a diuretic). Five of the individuals so treated have shown gratifying improvement in blood pressure and general health, resulting in their rehabilitation following previous incapacitation in 2 instances. However, 3 have experienced only slight reduction in blood pressure and 5 patients have noticed no improvement which seemed attributable to this modification in therapy, including 3 who have died in the course of the two-year period of follow-up: one with an enlarging abdominal aortic aneurysm, one after a massive myocardial infarct, and one with rapidly progressive renal insufficiency.

3) Four patients have had Cushing's syndrome which has been treated by adrenalectomy with an excellent result in 1 patient, by hypophysectomy with little improvement in a second, and by no surgical treatment because of severe concomitant renal insufficiency in the remaining 2 patients. One of these last 2 individuals has died from a ruptured aortic aneurysm, and the other has required renal transplantation.

4) In 8 patients (13%), the further diagnostic studies have not yet progressed to the point where the nature of any underlying renal disorder has been defined. These patients, therefore, are still unclassified.

5) In the remaining 19 patients (31%) there has been no evidence of renal parenchymal disease, no macroscopically evident unilateral or bilateral renal arterial stenosis, yet peripheral vein PRA levels were elevated and renal vein PRA levels were equal on the 2 sides. This combination of findings presumably resulted either from bilateral renal ischaemia conse-

quent upon obstruction of microscopical portions of the renal vascular tree, or from some relatively specific stimulus to hyperfunction of the juxtaglomerular apparatus. Further study of these patients is needed.

We conclude from these observations that the presence of a hypotensive response to saralasin infusions provides a simple, safe and useful means of discovering angiotensinogenic hypertension among large populations of hypertensive subjects. The procedure, as currently performed, is not infallible, being associated with approximately 5% of false-positive and occasional false-negative results, but is far better than measurement of stimulated PRA in peripheral blood.

Summary

Specific antagonists of angiotensin II (AII) such as saralasin might theoretically be of great value in the recognition of angiotensinogenic hypertension. Evidence is presented to show the importance of overcoming any existing sodium overload and of administering saralasin first in small and then in larger amounts by infusion (or injection). When this was done in 600 hypertensive patients, 62 showed a fall in blood pressure of more than 10/8 mm Hg. Further tests in 50 of these subjects indicated that the fall in blood pressure was associated with high peripheral levels of plasma renin activity (PRA) and/or abnormal renal vein PRA ratios in 94%. The procedure rarely failed to detect even mild forms of angiotensinogenic hypertension. In the 62 patients found to have angiotensinogenic hypertension, the responsible lesions included unilateral renal arterial stenosis with good contralateral renal function (29%), bilateral renal disease (21%), Cushing's syndrome (6%), small vessel disease or specific excess of renin production – without other detectable renal disease – (31%) and incompletely evaluated disorders (13%). Saralasin has been of great value in simply and reliably demonstrating the presence or absence of an angiotensinogenic component in a large group of hypertensive patients.

References

1 BARRACLOUGH, M. A.; BACCHUS, B.; BROWN, J. J.; DAVIES, D. L.; LEVER, A. F., and ROBERTSON, J. I. S.: Plasma-renin and aldosterone secretion in hypertensive patients with renal artery lesions. Lancet *ii:* 1310–1313 (1965).

2   COHEN, E. L.; ROVNER, D. R., and CONN, J. W.: Postural augmentation of plasma renin activity. Importance in diagnosis of renovascular hypertension. J. Am. med. Ass. *197:* 973–978 (1966).
3   BRUNNER, H. R.; GAVRAS, H., and LARAGH, J. H.: Angiotensin II blockade in man by sar^1-ala^8-angiotensin II for understanding and treatment of high blood pressure. Lancet *ii:* 1045–1048 (1973).
4   STREETEN, D. H. P.; ANDERSON, G. H., jr.; FREIBERG, J. M., and DALAKOS, T. G.: Use of an angiotensin II antagonist (saralasin) in the recognition of 'angiotensinogenic' hypertension. New Engl. J. Med. *292:* 657–662 (1975).
5   STREETEN, D. H. P.; FREIBERG, J. M.; ANDERSON, G. H., jr., and DALAKOS, T. G.: Identification of angiotensinogenic hypertension in man using 1-sar-8-ala-angiotensin II (saralasin, P-113). Circulation Res. *36–37:* suppl. I, pp. 125–132 (1975).
6   JUDSON, W. E. and HELMER, O. M.: Diagnostic and prognostic values of renin activity in renal venous plasma in renovascular hypertension. Hypertension. Proc. Council for High Blood Pressure Res., 1964, vol. 13, pp. 79–89.
7   MICHELAKIS, A. M.; FOSTER, J. H., and LIDDLE, G. W.: Measurement of renin in both renal veins – its use in diagnosis of renovascular hypertension. Archs intern. Med. *120:* 444–448 (1967).
8   GUNNELLS, J. C.; MCGUFFIN, W. L.; JOHNSRUDE, I., and ROBINSON, R. R.: Peripheral and renal venous plasma renin activity in hypertension. Ann. intern. Med. *71:* 555–575 (1969).
9   KAPLAN, N. M. and SILAH, J. G.: The angiotensin infusion test. A new approach to the differential diagnosis of renovascular hypertension. New Engl. J. Med. *271:* 536–541 (1964).
10 HOCKEN, A. G.; KARK, R. M., and PASSOVOY, M.: The angiotensin infusion test. Lancet *i:* 5–10 (1966).
11 WINER, B. M.; LUBBE, W. F., and SIMON, M.: Renin in the diagnosis of renovascular hypertension. J. Am. med. Ass. *202:* 121–128 (1967).
12 BROWN, T. C.; DAVIES, J. O.; OLICHNEY, M. J., and JOHNSTON, C. I.: Relation of plasma renin to sodium balance and arterial pressure in experimental renal hypertension. Circulation Res. *18:* 475–483 (1966).
13 MILLER, E. D.; SAMUELS, A. I.; HABER, E., and BARGER, A. C.: Inhibition of angiotensin conversion in experimental renovascular hypertension. Science *177:* 1108–1109 (1972).
14 GAVRAS, H.; BRUNNER, H. R.; VAUGHAN, E. D., jr., and LARAGH, J. H.: Angiotensin-sodium interaction in blood pressure maintenance of renal hypertensive and normotensive rats. Science *180:* 1369–1372 (1973).
15 GAVRAS, H.; BRUNNER, H. R.; THURSTON, H., and LARAGH, J. H.: Reciprocation of renin dependency with sodium volume dependency in renal hypertension. Science *188:* 1316–1317 (1975).
16 STREETEN, D. H. P.; ANDERSON, G. H., jr., and DALAKOS, T. G.: Angiotensin blockade: its clinical significance. Am. J. Med. *60:* 817–824 (1976).
17 PETTINGER, W. A. and MITCHELL, H. C.: Renin release, saralasin and vasodilator-beta-blocker drug interaction. New Engl. J. Med. *292:* 1214–1217 (1975).
18 WALLACE, J. M.; CASE, D. B., and LARAGH, J. H.: Immediate pressor response to saralasin (in press, 1976).

19 ANDERSON, G. H., jr.; FREIBERG, J. M.; DALAKOS, T. G., and STREETEN, D. H. P.: Agonistic response of vascular receptors to 1-sar-8-ala-angiotensin II (saralasin). Clin. Res. 23: 218A (1975).
20 MARKS, L. S.; MAXWELL, M. H., and KAUFMAN, J. J.: Saralasin bolus test: rapid screening procedure for renin-mediated hypertension. Lancet ii: 784–787 (1975).
21 STREETEN, D. H. P.; ANDERSON, G. H., jr.; FREIBERG, J. M., and DALAKOS, T. G.: Angiotensin II blockade in the hypertensive patient. Hospital Practice, 10: 83–90 (1975).
22 MACGREGOR, G. A. and DAWES, P. M.: Saralasin bolus test. Lancet ii: 923 (1975).

DAVID H. P. STREETEN, MB DPhil., Department of Medicine, State University of New York Upstate Medical Center, 750 East Adams Street, *Syracuse NY 13210* (USA)

Discussion[1]

Question from floor: Dr. STREETEN, how valid is the use of furosemide in preparing patients for a saralasin test and how much does individual sensitivity to the diuretic influence the subsequent response? In this regard, have you studied the effects of different doses of furosemide?

STREETEN: Preparing patients for a study of their blood pressure responses to saralasin by sodium depletion is very laborious. You cannot study very many subjects that way, and that is its main limitation; however, it is preferable to the use of a diuretic, I believe. One could probably accomplish the same thing with a slightly smaller dose of furosemide than we have used. We did not do a dose-response curve of the effects of various doses of furosemide. It did not seem to be worthwhile when we found that the 40-mg dose worked well.

I believe that the question of how much diuresis one should induce is very important, though, because it is possible that if you deplete an individual too severely, you might, as it were, induce a response to saralasin. We have not yet seen this in any of our subjects, but theoretically it could happen with larger doses.

PETTINGER: The question is a tremendously important one – I think it relates to the overall concept of what diagnostic tools are about. It seems to me that we need greater precision in determining the frequency of false-positives and false-negatives, but particularly, the false-negative test with saralasin, if we are to use this peptide in screening for renal artery stenosis. The only way that I know of to exclude renal artery stenosis is by arteriography. Arteriograms in the States now are not accepted as a routine for exclusion of renal artery stenosis in the hypertensive population at large. Thus, there is no way for us to really find out the level of accuracy of this testing with saralasin except possibly in Centers like Vanderbilt where, for other reasons, they are doing this type of study. Secondly, we need to know the dose-response curve of the diuretic-induction of saralasin response in patients with low renin hypertension and those with normal renin hypertension and those with renal artery stenosis. This type of information is a pre-requisite for the experimental

1   Discussion of papers of PETTINGER and MITCHELL and STREETEN *et al.*

design that we need here to clearly establish the incidence of false-positives and negatives. Dr. STREETEN, what is your reaction to that?

STREETEN: I am not sure, but I think the incidence of the false-negative response to saralasin is extremely low. We should be prepared to watch the potential candidates for this phenomenon if, in fact, it exists at all, which I expect it does.

OELKERS: Dr. PETTINGER, in your scheme you suggest that the effect of saralasin on renin release is mediated by the blocking of angiotensin feedback, but you did not show the changes in blood pressure in your experiments. I think the very fact that propranolol blocks the increase of renin could instead mean that this is a counter-regulation of the autonomic nervous system.

PETTINGER: That is the critical point, I certainly concur. Dr. KEETON, Dr. MITCHELL and I have done extensive studies in man and in the rat [Circulation Res. in press June-July 1976]. Renin release with saralasin occurs in the absence of blood pressure changes in states of normal sodium balance. Also, equidepressor doses of hydralazine and saralasin in salt-depleted rats result in 100-fold greater renin release with saralasin than with hydralazine. These results suggest a selective receptor effect of saralasin. There is some baroreceptor stimulation to renin release when blood pressure is lowered with saralasin but this is not the entire explanation for it.

MULROW: Dr. PETTINGER, I would interpret your data a little differently. The greater response in PRA with saralasin after sodium depletion may be attributed to the fact that volume depletion enhances the renin-release response to a lot of stimuli. Now, it may be that you were shifting the dose-response curve of the agonist effect of saralasin on renin-release to the left. If you look at your dose-response data on renin-release in the rat on the normal sodium diet, at huge doses where you are clearly blocking any circulating angiotensin II effect there was an agonist effect of saralasin on the plasma-renin activity. What you have done with volume depletion is shifted the dose-response curve over, and now you are looking at an agonist effect, and not necessarily a block of the negative feedback.

PETTINGER: I could extend the generalization further, Dr. MULROW, in the following way. The quantity of renin release induced by any stimulatory intervention is determined by the activity at the time of the intervention.

SALVETTI: Dr. STREETEN, I wonder why you must deplete your renovascular patients to show the role of the renin-angiotensin system in causing high blood pressure. If you measure plasma-renin activity in renal veins in advance, you find that only the ischaemic kidney contributes to the peripheral levels of plasma renin activity, and if you suppress the renin secretion from the ischaemic kidney either by mononephrectomy, or by revascularization, blood pressure is normalized. So if you block the renin-angiotensin system under conditions of normal sodium intake, you should decrease blood pressure, if this system is responsible for the genesis of the hypertension.

STREETEN: We attempt to precede saralasin studies with some form of natriuresis because there is a lot of evidence in the literature that volume overload consequent upon angiotensin II excess becomes an important contributor to the hypertension and to maintaining it. Thus, one completely fails to detect the angiotensin component of hypertension in a patient unless one gets rid of the sodium and water overload, as both our studies in humans and the studies of LARAGH's group in animals have shown.

Discussion

STOKES: This is to answer Dr. OELKER's question of Dr. PETTINGER regarding renin release being dependent on the fall in blood pressure after blocker. We reported some results at the last Milan conference, in which we showed both in the conscious rabbit and in the vagotomised, atropinised and ansolysen-treated rat that we could demonstrate an increased renin release in the absence of blood pressure changes, moreover, in the absence of changes suggesting buffer-mediated effects.

In STOKES and EDWARDS: Drugs Affecting the Renin-Angiotensin-Aldosterone System. Use of Angiotensin Inhibitors

The Effects of the Angiotensin II Antagonist Saralasin on Blood Pressure and Plasma Aldosterone in Man in Relation to the Prevailing Plasma Angiotensin II Concentration

J. J. BROWN, W. C. B. BROWN, R. FRASER, A. F. LEVER, J. J. MORTON, J. I. S. ROBERTSON, E. A. ROSEI and P. M. TRUST

Medical Research Council Blood Pressure Unit, Western Infirmary, Glasgow

Introduction	230
Methods	231
Results	233
Normal Subjects	233
Essential Hypertension	234
Primary Aldosteronism	234
Hypertension with Severe Chronic Renal Failure	234
Hypertension with Unilateral Renal Artery Stenosis	236
Relationship between Basal Plasma Angiotensin II Concentration and Blood Pressure Change on Saralasin Infusion	236
Relationship between Basal Plasma Angiotensin II concentration and Change in Plasma Aldosterone on Saralasin Infusion	237
Discussion	237
Summary	239
Acknowledgements	240
References	240

Introduction

In recent years, the combination of radioimmunoassay for the measurement of plasma angiotensin II concentration with the infusion of synthetic angiotensin II has permitted more accurate quantitative assessment of the renin-angiotensin system than was possible hitherto. Dose-response curves of the relationship of angiotensin II to arterial pressure and plasma aldosterone concentration have been constructed in a variety of physiological and pathological states [1-6]. These studies have shown that sodium depletion in normal subjects enhances the aldosterone-stimulant effect, while simultaneously diminishing the pressor effect of angiotensin II. In untreated renal and malignant hypertension, both the pressor and aldosterone-stimulant effects of angiotensin II are enhanced [7-9].

Further insight into these problems has been attained with the availability of substances which specifically block the actions of certain components of the renin-angiotensin system [10]. Amongst these is saralasin (Morton-Norwich Laboratories), an angiotensin II analogue (Sar^1-Ala^8) synthesized by PALS et al. [11], which specifically antagonizes angiotensin II.

We report here the effects of infusions of saralasin in normal subjects and in hypertensives of varied aetiology in different states of sodium balance. Particular attention was paid to the effects of saralasin in relation to the basal level of angiotensin II.

Methods

Details of the methods used have been given in an earlier paper by BROWN et al. [12]. The experimental protocol is shown in figure 1. Throughout the saralasin infusion, and for at least the preceding hour, the subjects remained supine. Infusions were given at a constant rate by an electrically driven pump (Dascon, Unden) via a plastic cannula into a fore-

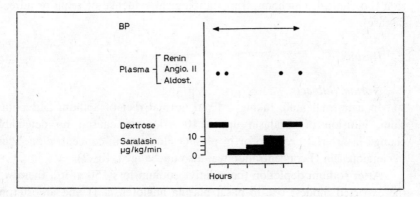

Fig. 1. Diagram of experimental saralasin infusion protocol.

arm vein. Blood samples were taken from a vein in the opposite forearm (fig. 1) via an indwelling plastic cannula for measurements of plasma concentrations of renin [13], angiotensin II [14] and aldosterone [15]. No cross-reaction between angiotensin II and saralasin was found in the assay system within the range 10–10,000 pg; however, there was a detectable cross-reaction with saralasin over 20,000 pg which, at the maximum rate of infusion (10/µg/kg/min), could be equivalent to an apparent plasma angiotensin II increment of up to 20 pg/ml.

Arterial blood pressure was measured indirectly every 3–5 mins by a semi-automatic machine (Elag, Köln), checked at intervals by a traditional clinical sphygmomanometer applied to the opposite forearm. In most cases, the diet was fixed to contain known and normal amounts of sodium and potassium for at least 3 days before the infusions. Sodium depletion was produced by reducing sodium intake to 10 mEq/day for 3 days and by frusemide (40–80 mg) given by mouth once on the first day of the reduced intake. None of the patients had taken diuretics, potassium supplements, oral contraceptives or hypotensive drugs (other than bethanidine in a few instances), for at least 4 weeks before infusion. In the severely hypertensive patients treatment with bethanidine was continued until the day before infusion.

The following subjects were studied: (a) Two normal subjects before and after sodium depletion. (b) One patient from whom a typical aldosterone-secreting adenoma was later removed. (c) 21 patients with hypertension and unilateral or bilateral renal disease (4 undergoing regular haemodialysis); 11 of these were infused with saralasin both before and after sodium depletion. (d) Three patients with apparently essential hypertension (no clinical, biochemical or radiological evidence of renal or adrenal disease).

Results

Normal Subjects
In a normal male taking a fixed normal diet of sodium and potassium, infusion of saralasin up to 10 µg/kg/min caused no detectable change in arterial pressure or in plasma aldosterone concentration. Plasma angiotensin II concentration was also unchanged (fig. 2).

After sodium depletion (cumulative sodium loss 150 mEq), there was the expected modest rise in basal plasma angiotensin II and aldosterone concentrations. Infusion of saralasin then led to a distinct fall in circulat-

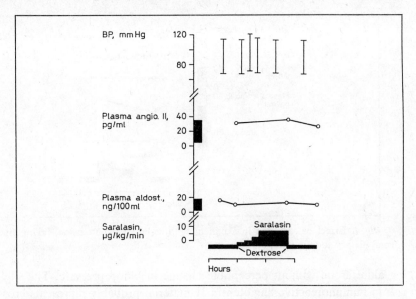

Fig. 2. Effect of saralasin infusion on blood pressure, plasma angiotensin II and plasma aldosterone in a normal subject taking fixed normal dietary sodium and potassium.

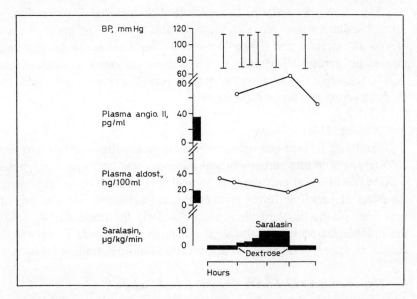

Fig. 3. Effect of saralasin infusion on blood pressure, plasma angiotensin II and plasma aldosterone in a normal subject after sodium depletion.

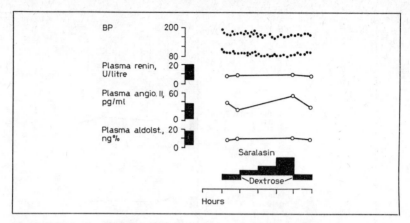

Fig. 4. Effect of saralasin in a patient with essential hypertension, taking fixed normal dietary sodium and potassium.

ing aldosterone without perceptible change in blood pressure. The slight rise in immunoreactive angiotensin II material probably represents cross-reaction of the antiserum with saralasin at high rates of infusion (fig. 3).

Essential Hypertension

The effect of saralasin in a patient with essential hypertension, in normal sodium status and with normal basal levels of renin, angiotensin II and aldosterone, are shown in figure 4. Saralasin up to 10 µg/kg/min caused no perceptible change in blood pressure, renin or aldosterone. Again, the slight rise in apparent angiotensin II concentration could well be due to cross-reaction with saralasin.

Primary Aldosteronism

Saralasin, infused pre-operatively in a patient from whom a typical aldosterone-secreting adreno-cortical adenoma was later removed, caused a slight rise in blood pressure (fig. 5), probably because of a slight agonistic effect at very low levels of endogenous angiotensin II. The already very high plasma level of aldosterone similarly increased slightly, while the low plasma renin concentration remained unchanged. There was, as before, an apparent slight rise in angiotensin II during saralasin infusion.

Hypertension with Severe Chronic Renal Failure

In a patient with severe hypertension and chronic renal failure undergoing regular haemodialysis (fig. 6), a modest infusion of saralasin

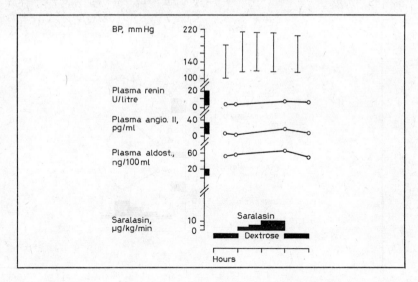

Fig. 5. Effect of saralasin in a patient with an aldosterone-secreting adrenocortical adenoma.

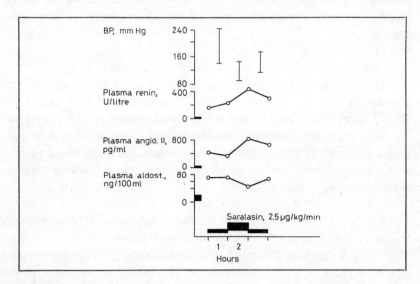

Fig. 6. Effect of saralasin in a patient with severe hypertension and chronic renal failure undergoing regular haemodialysis.

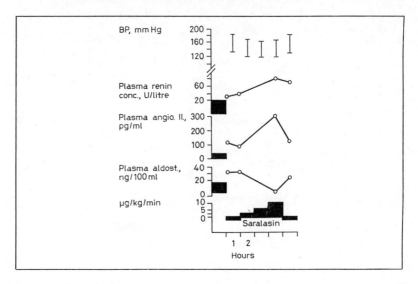

Fig. 7. Effect of saralasin in a patient with hypertension and unilateral renal artery stenosis.

(2.5 µg/kg/min) caused a marked fall in arterial pressure. The very high basal levels of renin and angiotensin II increased further, while the elevated plasma aldosterone concentration fell, although not into the normal range.

Hypertension with Unilateral Renal Artery Stenosis

In a patient with hypertension, unilateral renal artery stenosis, and abnormally high basal levels of renin, angiotensin II and aldosterone (fig. 7), saralasin induced a clear fall in blood pressure, and restored plasma aldosterone concentration into the normal range, while plasma renin and angiotensin II concentrations rose. These effects on blood pressure and aldosterone were enhanced when saralasin was infused in renal hypertension after sodium depletion.

Relationship between Basal Plasma Angiotensin II Concentration and Blood Pressure Change on Saralasin Infusion

Figure 8 shows, in the entire hypertensive series, the change in systolic arterial pressure on saralasin infusion plotted against the basal plasma angiotensin II concentration. There is a close inverse correlation between the two measurements ($r = -0.86$; $p<0.001$), the blood pressure fall

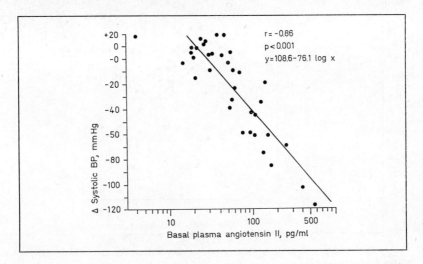

Fig. 8. Entire hypertensive series, showing change in systolic arterial pressure on saralasin administration in relation to the basal (pre-saralasin) plasma angiotensin II level.

being greatest where the basal plasma angiotensin II was high and being negligible (in some instances an increase in pressure was seen) when basal plasma angiotensin II concentration was low.

Relationship between Basal Plasma Angiotensin II Concentration and Change in Plasma Aldosterone on Saralasin Infusion

Figure 9 shows a similar plot of the change in plasma aldosterone concentration on saralasin infusion against the basal plasma angiotensin II concentration. This relationship is remarkably similar to that relating blood pressure to plasma angiotensin II, the fall in aldosterone being greatest where basal angiotensin II was highest. Though the correlation is less close than in the case of the blood pressure change, it remains highly significant ($r = -0.52$; $p < 0.001$).

Discussion

The present results confirm and amplify dose-response data previously obtained from angiotensin II infusions combined with plasma angiotensin II and aldosterone assays [1–9]. It has been shown previously

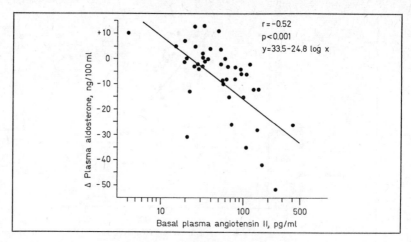

Fig. 9. Change in plasma aldosterone concentration on saralasin infusion in relation to the basal (pre-saralasin) plasma angiotensin II level.

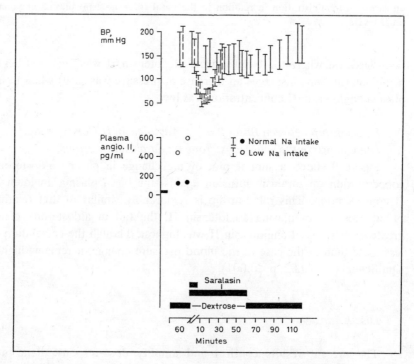

Fig. 10. Patient with renal hypertension, showing enhanced effects of saralasin after sodium depletion.

that while, in normal subjects, plasma angiotensin II concentrations are within a range having an effect on arterial pressure [16], the values are at the lower end of a sigmoid dose-response curve. Thus, variations of angiotensin II within the normal range affected blood pressure only by 5–10 mm Hg [2, 17]. Therefore, eliminating the pressor effect of angiotensin II in this range, as in the present experiments, has a negligible effect on blood pressure.

Sodium depletion in normal subjects shifts the pressor dose-response curve in a roughly parallel fashion to the right [2]; the prevailing angiotensin II levels, though increased, remain at the lower end of sigmoid dose-response curves and again, the effect of blocking angiotensin II, as in the present study, does not affect blood pressure greatly (fig. 3).

However, in renal hypertension the pressor effect of a given plasma angiotensin II concentration is enhanced [7–9] and in these circumstances, when saralasin is given, blood pressure falls steeply (fig. 10).

The angiotensin II/aldosterone dose-response relationships are somewhat different, as is emphasized in the present studies. In normal sodium-replete man, the dose-response curve is shallow [1, 2, 4]; thus, when saralasin is infused, little change in plasma aldosterone is seen. When sodium is removed from normal subjects, however, the angiotensin II/aldosterone dose-response curve is steepened [1, 2, 4] and, in these circumstances, when saralasin is infused, a distinct fall in plasma aldosterone is seen in contrast to the negligible effect on blood pressure (fig. 3).

In renal hypertension, the angiotensin II/aldosterone dose-response curve is also steepened [7–9], and saralasin causes a sharp fall in plasma aldosterone concentration as in blood pressure (fig. 7). These effects are enhanced if a patient with renal hypertension is also sodium-depleted (fig. 10).

In addition to confirming these dose-response relationships, the present study underlies the importance of the renin-angiotensin system in the stimulation of aldosterone in sodium-deplete normal man, and in maintaining the elevation of both arterial pressure and plasma aldosterone in renal hypertension.

Summary

The effect of saralasin in lowering blood pressure and plasma aldosterone concentration in normal subjects, both sodium-replete and sodium-deplete, and in patients with various forms of hypertension, is closely related to the basal plasma angiotensin II concentration. These findings confirm and extend earlier studies of angiotensin II/arterial pressure and angiotensin II/aldosterone dose-response curves. They also emphasize the importance of the renin-angiotensin system in the control of aldosterone in sodium depletion and in renal hypertension.

Acknowledgements

We wish to thank Drs. R. KEENAN, M. HARRIES and G. HOOPER and Mrs. M. VERO of Eaton Laboratories Ltd. for generously supplying the saralasin, and Drs. A. ROBERTSON and D. BARLOW for their invaluable help with the studies in normal subjects.

References

1 BROWN, J. J.; FRASER, R.; LEVER, A. F.; MORTON, J. J.; OELKERS, W.; ROBERTSON, J. I. S., and YOUNG, J.: in SAMBHI Mechanisms of hypertension, pp. 148–154 (Excerpta Medica, Amsterdam 1973).
2 OELKERS, W.; BROWN, J. J.; FRASER, R.; LEVER, A. F.; MORTON, J. J., and ROBERTSON, J. I. S.: Sensitisation of the adrenal cortex to angiotensin II in sodium-deplete man. Circulation Res. *34:* 69–77 (1974).
3 OELKERS, W.; SCHONESHOFER, M.; SCHULTZE, G.; BROWN, J. J.; FRASER, R.; LEVER, A. F.; MORTON, J. J., and ROBERTSON, J. I. S.: Effect of prolonged low dose angiotensin II infusion on the sensitivity of the adrenal cortex in man. Circulation Res. *36–37:* suppl. I, pp. 49–56 (1975).
4 HOLLENBERG, N. K.; CHENITZ, W. R.; ADAMS, D. F., and WILLIAMS, G. H.: Reciprocal influence of salt intake on adrenal glomerulosa and renal vascular responses to angiotensin II in normal man. J. clin. Invest. *54:* 34–42 (1974).
5 HOLLENBERG, N. K.; WILLIAMS, G.; BURGER, B., and HOOSHMAND, I.: The influence of potassium on the renal vasculature and the adrenal gland, and their responsiveness to angiotensin II in normal man. Clin. Sci. mol. Med. *49:* 527–534 (1975).
6 DEHENEFFE, J. ; CUESTA, V.; BRIGGS, J. D.; BROWN, J. J.; FRASER, R.; LEVER, A. F.; MORTON, J. J.; ROBERTSON, J. I. S., and TREE, M.: Response of aldosterone and blood pressure to angiotensin II infusion in anephric man: effect of sodium deprivation. Circulation Res. (in press).
7 BEEVERS, D. G.; BROWN, J. J.; CUESTA, V.; FRASER, R.; LEVER, A. F.; MORTON, J. J.; OELKERS, W., and ROBERTSON, J. I. S.: The role of angiotensin II in the control of aldosterone and blood pressure in normal and hypertensive man. Proc. 6th Int. Study Crp. Steroid Hormones, Rome 1973, pp. 291–300 (Pergamon Press, Oxford 1973).

8   BEEVERS, D. G.; BROWN, J. J.; CUESTA, V.; DAVIES, D. L.; FRASER, R.; LEBEL, M.; LEVER, A. F.; MORTON, J. J.; OELKERS, W.; ROBERTSON, J. I. S.; SCHALEKAMP, M. A., and TREE, M.: Inter-relationships between plasma angiotensin II, arterial pressure, aldosterone and exchangeable sodium in normotensive and hypertensive man. J. Steroid Biochem. 6: 779–784 (1975).

9   BEEVERS, D. G.; BROWN, J. J.; FRASER, R.; LEVER, A. F.; MORTON, J. J.; ROBERTSON, J. I. S.; SEMPLE, P. F., and TREE, M.: The clinical value of renin and angiotensin estimations. Kidney int. 8: suppl., pp. 181–201 (1975).

10 DAVIS, J. O.: The use of blocking agents to define the functions of the renin-angiotensin system. Clin. Sci. mol. Med. 48, Suppl. 2: 3–14 (1975).

11 PALS, D. T.; MASUCCI, F. D.; DENNING, G. S.; SIPOS, F., and FESSLER, D. C.: Role of the pressor action of angiotensin II in experimental hypertension. Circulation Res. 29: 673–681 (1971).

12 BROWN, J. J.; BROWN, W. C. B.; FRASER, R.; LEVER, A. F.; MORTON, J. J.; ROBERTSON, J. I. S.; ROSEI, E. A.; TREE, M., and TRUST, P. M.: The effects of infusing the angiotensin II antagonist saralasin on arterial blood pressure and on the plasma concentrations of aldosterone, renin and angiotensin II in normal subjects and selected hypertensive patients: in SAMBHI The Effects of Antihypertensive Therapy (Excerpta Medica, Amsterdam, in press).

13 BROWN, J. J.; DAVIES, D. L.; LEVER, A. F.; ROBERTSON, J. I. S., and TREE, M.: The estimation of renin in human plasma. Biochem. J. 93: 594–600 (1964).

14 DÜSTERDIECK, G. and MCELWEE, G.: Estimation of angiotensin II concentration in human plasma by radioimmunoassay. Some applications to physiological and clinical states. Eur. J. clin. Invest. 2: 32–38 (1971).

15 FRASER, R.; GUEST, S., and YOUNG, J.: A comparison of double isotope derivative and radioimmunological estimation of plasma aldosterone concentration in man. Clin. Sci. mol. Med. 45: 411–415 (1973).

16 CHINN, R. H. and DÜSTERDIECK, G.: The response of blood pressure to infusion of angiotensin II; relation to plasma concentrations of renin and angiotensin II. Clin. Sci. 42: 489–504 (1972).

17 OELKERS, W.; DÜSTERDIECK, G., and MORTON, J. J.: Arterial angiotensin II and venous immunoreactive material before and during angiotensin infusion in man. Clin. Sci. 43: 209–218 (1972).

Dr. J. J. BROWN, Medical Research Council Blood Pressure Unit, Western Infirmary, *Glasgow G11 6NT* (Scotland)

Haemodynamic Effects of Sar[1]-Ala[8]-Angiotensin II in Patients with Renovascular Hypertension

ROBERT FAGARD, ANTOON AMERY, PAUL LIJNEN, TONY REYBROUCK and LEON BILLIET

Department of Medicine, Cardiopulmonary Laboratory and Laboratory for Hypertension, University of Leuven, Leuven

Contents

Introduction	242
Methods	243
Results	244
Characteristics of the Patients	244
Mean Arterial Pressure and Heart Rate in Sodium-Replete Patients	245
Mean Arterial Pressure and Haemodynamic Variables in Sodium-Deplete Patients	245
Mean Arterial Pressure in Patients Studied before and after Sodium Depletion	246
Discussion	247
Summary	248
References	248

Introduction

Saralasin (Sar[1]-Ala[8]-angiotensin II) is a specific antagonist of the vascular effects of angiotensin II [1]. In the present study, saralasin was used to investigate the angiotensin dependency of renovascular hypertension in man. Recent studies on the identification of angiotensinogenic hypertension using saralasin were performed after patients had been sodium-depleted [2–4]. This study reports on the effects of saralasin in patients with the probable diagnosis of renovascular hypertension in both the sodium-replete and the sodium-deplete states.

Table I. Characteristics of patients

Age (years; mean ± SD)	39.4 ± 11.2
Sex	
Male (n)	8
Female (n)	6
Weight (kg; mean ± SD)	64.5 ± 6.9
Recumbent blood pressure on admission to hospital	192.4 ± 20.4
(mm Hg; mean ± SD)	119.7 ± 11.6
Eye fundus	
Normal (n)	2
Grade 1 (n)	7
Grade 2 (n)	3
Grade 3 (n)	2
Grade 4 (n)	0
LVH on ECG (n)	5
Creatinine clearance (ml/min/1.73 m^2; mean ± SD)	82.9 ± 22.1

Methods

The patients studied were 14 hypertensive adults in whom antihypertensive treatment had been interrupted for at least 3 weeks. Routine studies were performed including a rapid sequence intravenous pyelogram and renal arteriogram. Renal vein blood and femoral artery blood were collected for determination of plasma renin concentration (PRC) after at least 3 days of sodium depletion by a low sodium diet and chlorthalidone 50 mg/day.

Saralasin[1] was infused intravenously at a rate of 10 µg/kg/min in six patients who were on a daily sodium intake of 130 mEq and in twelve patients after at least 3 days of sodium depletion by diet and chlorthalidone. The tests were performed in the recumbent position, in the morning after a light breakfast. A catheter (Vygon, 115.09) was introduced in the brachial artery for recording the intra-arterial pressure and in 9 sodium-depleted patients a Swan-Ganz flow-directed catheter (Edwards Lab., 93-110-5-F) was positioned in the pulmonary artery. Heart rate was monitored continuously. Cardiac output was determined by the oxygen-Fick method. For measurement of oxygen-uptake, the open-circuit method was used and oxygen was determined by a paramagnetic gas analyzer. Oxygen

1   Generously supplied by Norwich Pharmacal Company.

content in arterial and mixed venous blood was calculated by the product of haemoglobin concentration (cyanmethaemoglobin method) and oxygen saturation (reflection oxymetry) times 1.34. A control period of at least 45 min was observed while glucose 5% was infused intravenously. Then the infusion of saralasin was started for 30 min in 5 sodium-depleted patients and for 60 min in all other tests. Patients were observed for 30 or 60 min after interruption of the drug administration. Cardiac output was determined at the end of the control period, 10, 30 and 60 min after the start of saralasin and at the same intervals after its interruption. PRC was measured at the end of the control period, using the method of SKINNER [5] (normal range, sodium-replete: 8–25 U/ml). The two-tailed Student's t-test was used for statistical analysis.

Results

Characteristics of the Patients

Some of the characteristics of the patients are given in table I. The arteriogram revealed a unilateral renal artery stenosis in 11 patients – 10 were arteriosclerotic lesions and 1 fibromuscular dysplasia. Of the other 3 patients, 1 had bilateral fibromuscular dysplasia and 2 had a renal artery aneurysm. Renal vein blood sampling was performed in 10 patients and showed a renal vein/aortic PRC ratio greater than 1.5 on the 'diseased' side in 7 and less than 1.5 in 3 patients. In the latter 3 patients and in the 4 in whom renal vein blood was not collected, the peripheral PRC level was high.

Mean Arterial Pressure and Heart Rate in
Sodium-Peplete Patients

Of the six patients studied in the sodium-replete state, five had stenosis of the renal artery with renal vein/arterial PRC ratios greater than 1.5 on the stenotic side in 4. Peripheral PRC was elevated in the 5th and also in the 6th patient who had a renal arterial aneurysm. As shown in table II, mean arterial pressure remained unchanged when saralasin was infused for 60 min in these patients. The maximal decrease in pressure was 2 mm Hg. In 4 of the 6 patients, the infusion rate was increased to 100 µg/kg/min for 30 min which resulted in an increase in mean arterial pressure of 5.0 ± 2.6 (SD) mm Hg ($p<0.05$). Heart rate remained unchanged throughout the test.

Table II. Mean arterial pressure and heart rate in six sodium-replete patients before, during and after the infusion of saralasin (10 µg/kg/min; mean ± SD)

	Control 0 min	Saralasin 10 min	30 min	60 min	Recovery 30 min
Mean arterial pressure	122.7 mm Hg ±8.3	100.7% ±4.0	100.4% ±2.7	101.0% ±2.2	102.4% ±5.2
p vs control	–	>0.1	>0.1	>0.1	>0.1
Heart rate	72.8 ±6.8	100.4% ±4.2	101.8% ±3.8	98.4% ±5.8	97.2% ±6.5
p vs control	–	>0.1	>0.1	>0.1	>0.1

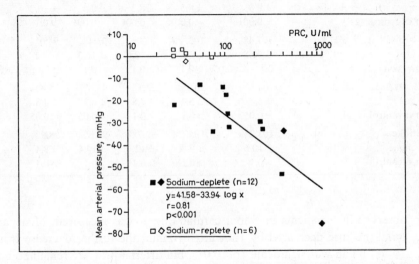

Fig. 1. Relationship between the log of plasma renin concentration and the change in mean arterial pressure during Sar^1-Ala^8-angiotensin II infusion in patients with renovascular hypertension in the sodium-replete state and after sodium depletion (sodium-replete: □ = renal artery stenosis, ◊ = renal artery aneurysm; sodium-deplete: ■ = stenosis, ♦ = aneurysm).

Mean Arterial Pressure and Haemodynamic Variables in Sodium-Deplete Patients

In the 12 patients studied in the sodium-deplete state, the decrease in mean arterial pressure ranged from 13 to 76 mm Hg and was closely related to the log PRC (fig. 1). Arterial pressure reached 100 mm Hg or less in 9 of the 12 patients. Table III shows the haemodynamic changes

Table III. Mean arterial pressure and haemodynamic variables in nine sodium-deplete patients before, during and after the infusion of saralasin (10 µg/kg/min; means ± SD)

	Control 0 min	Saralasin 10 min	30 min	60 min	Recovery 10 min	30 min	60 min
Number of observations	9	9	9	7	9	9	7
Mean arterial pressure	130.4 mm Hg ±18.8	88.1% ±5.9	76.4% ±12.2	74.3% ±14.9	76.0% ±14.9	85.8% ±16.5	96.3% ±12.7
p vs control	–	<0.001	<0.001	<0.01	<0.01	<0.05	>0.1
Total peripheral resistance	21.7 ±4.4	82.4% ±8.8	72.4% ±8.5	78.9% ±8.2	81.1% ±8.4	95.4% ±10.2	103.6% ±13.3
p vs control	–	<0.001	<0.001	<0.001	<0.001	>0.1	>0.1
Cardiac output	6.21 l/min ±1.38	107.4% ±10.8	103.3% ±10.9	93.9% ±12.6	93.4% ±16.3	94.0% ±8.2	92.9% ±6.3
p vs control	–	0.1>p>0.05	>0.1	>0.1	>0.1	0.1>p>0.05	<0.05
Heart rate	91.8 ±12.7	104.6% ±5.4	98.3% ±5.7	94.1% ±7.8	93.8% ±6.3	95.8% ±4.8	95.9% ±4.1
p vs control	–	<0.05	>0.1	=0.1	<0.05	<0.05	<0.05
Stroke volume	68.8 ml ±18.6	103.2% ±10.6	105.1% ±9.3	99.7% ±12.4	99.8% ±14.4	98.9% ±9.4	97.4% ±4.5
p vs control	–	>0.1	>0.1	>0.1	>0.1	>0.1	>0.1

observed in 9 patients in whom cardiac output was measured. Mean arterial pressure decreased by 12% after 10 min; the further decrease from 10 to 30 min was significant (p<0.01), but arterial pressure remained unchanged when the infusion of saralasin was continued from 30 to 60 min. Slight increases in heart rate and cardiac output were observed 10 min after the start of the infusion of saralasin but both variables decreased afterwards and remained somewhat below the control values during the recovery period. Stroke volume remained essentially unchanged throughout the test. Saralasin caused a clear decrease in calculated total peripheral resistance.

Mean Arterial Pressure in Patients Studied before and after Sodium Depletion

In 4 patients in whom saralasin was infused in both the sodium-replete and the sodium-deplete state, the drug caused no change in mean arterial pressure in the sodium-replete state (126.7 ± 12.6 vs a control level of $126.2 \pm SD$ 9.5 mm Hg) ($p>0.1$), but arterial pressure decreased from 111.7 ± 13.4 to 80.5 ± 11.7 mm Hg after sodium depletion ($p<0.001$).

Discussion

The effects of the angiotensin II antagonist Sar^1-Ala^8-angiotensin II were studied in 14 patients with a probable diagnosis of hypertension of renovascular origin. When the drug was infused at a rate of 10 μg/kg/min in patients on a daily sodium intake of 130 mEq, mean arterial pressure remained unchanged although PRC was abnormally high. The 'saralasin test' for the detection of angiotensinogenic hypertension is generally performed after sodium depletion [2–4]. One can only speculate why the angiotensin II antagonist is ineffective in the sodium-replete state. The drug has some agonistic effect and an increase of mean arterial pressure of 5 mm Hg was observed when the infusion rate was increased to 100 μg/kg/min. It is theoretically possible that the antagonistic and the agonistic effects of saralasin are exactly balanced in sodium-replete patients. It is, however, more likely that chronic human renovascular hypertension is not dependent on the arteriolar effects of angiotensin II in the sodium-replete state. It could be sodium volume-dependent by virtue of increases in renal sodium reabsorption via increases in aldosterone secretion, and/or via such intrarenal physical factors as decreased glomerular filtration rate and increased proximal tubular reabsorption induced by the reduction in renal perfusion pressure and blood flow. In such circumstances, saralasin does not lead to a decrease in arterial pressure. An experimental counterpart of human unilateral renovascular hypertension, which was the case in most of the studied patients, is two-kidney renovascular hypertension in animals where one renal artery is clamped and the opposite kidney is left in place. In contrast to the observations in man, saralasin decreases blood pressure uniformly in such sodium-replete animals according to some authors [6] and to a variable degree according to others [7, 8].

When patients were sodium volume-depleted by a low sodium diet and a diuretic, which resulted in an average weight loss of 3.4 kg, saralasin caused a fast drop in arterial pressure and in total peripheral resistance, which indicates that the high blood pressure depends on the arteriolar vasoconstrictive effect of angiotensin II in these circumstances. Moreover, the correlation between the control PRC and the decrease in mean arterial pressure was highly significant (fig. 1). Combined sodium-volume depletion and inhibition of the arteriolar effects of angiotensin II resulted in a normalization of blood pressure in 9 of the 12 patients.

The haemodynamic changes induced by saralasin in sodium-deplete patients show that the decrease in arterial pressure was due to changes in total peripheral resistance probably through arteriolar dilatation. The drop in pressure led to a slight reflex increase in heart rate and in cardiac output after 10 min, but neither variable remained elevated during the continuation of the saralasin infusion, although blood pressure continued to fall. This contrasts with observations using potent arteriolar vasodilators, such as diazoxide, where the hypotension is associated with intense reflex increases in heart rate and cardiac output [9]. Possibly the reflex increase in heart rate is counterbalanced by inhibition of an angiotensin II-induced increase in heart rate. Indeed, angiotensin-induced tachycardia mediated via peripheral sympathetic nerves has been demonstrated [10].

Summary

Sar^1-Ala^8-angiotensin II was infused intravenously (10 μg/kg/min) in 14 patients with renovascular hypertension, including 11 with renal artery stenosis. Brachial artery pressure and heart rate remained unchanged in six patients who were on a daily sodium intake of 130 mEq. In 12 tests performed after sodium depletion, the decrease in mean arterial pressure ranged from 13 to 76 mm Hg and showed a significant correlation with the plasma renin concentration prevailing immediately before the infusion of the drug (r = 0.81; p<0.001). The hypotensive response was due to a drop in total peripheral resistance. Heart rate and cardiac output showed slight increases 10 min after the start of saralasin infusion.

References

1 PALS, D. T.; MASUCCI, F. D.; SIPOS, F., and DENNING, G. S.: A specific competitive antagonist of the vascular action of angiotensin II. Circulation Res. 29: 664–672 (1971).
2 STREETEN, D. H.; ANDERSON, G. H.; FREIBERG, J. M., and DALAKOS, T. G.: Use of an angiotensin II antagonist (saralasin) in the recognition of angiotensinogenic hypertension. New Engl. J. Med. 292: 657–662 (1975).
3 STREETEN, D. H.; FREIBERG, J. M.; ANDERSON, G. H., and DALAKOS, T. G.:

Identification of angiotensinogenic hypertension in man using 1-sar-8-ala-angiotensin II. Circulation Res. *36–37:* suppl. I, pp. 125–132 (1975).

4 MARKS, L. S.; MAXWELL, M. H., and KAUFMAN, J. J.: Saralasin bolus test. Lancet *ii:* 784–787 (1975).

5 SKINNER, S. L.: Improved assay methods for renin concentration and activity in human plasma. Circulation Res. *20:* 391–402 (1967).

6 BRUNNER, H. R.; KIRSHMANN, J. D.; SEALY, J. E., and LARAGH, J. H.: Hypertension of renal origin. Evidence for two different mechanisms. Science *174:* 1344–1346 (1971).

7 PALS, D. T.; MASUCCI, F. D.; DENNING, G. S.; SIPOS, F., and FESSLER, D. C.: Role of the pressor action of angiotensin II in experimental hypertension. Circulation Res. *29:* 673–681 (1971).

8 MCDONALD, G. J.; BOYD, G. W., and PEART, W. S.: Effect of the angiotensin II blocker 1-sar-8-ala-angiotensin II on renal artery clip hypertension in the rat. Circulation Res. *37:* 640–646 (1975).

9 KOCH-WESER, J.: Vasodilator drugs in the treatment of hypertension. Archs intern. Med. *133:* 1017–1027 (1974).

10 BUNAG, R. D.: Circulatory effects of angiotensin; in Angiotensin, p. 443 (Springer, Berlin 1974).

Dr. R. FAGARD, Cardiopulmonary Laboratory, Akademisch Ziekenhuis, Weligerveld 1, *B–3041 Pellenberg* (Belgium)

Discussion[1]

GEYSKES: Dr. ROBERTSON, since you found such a beautiful correlation between angiotensin II concentrations and the blood pressure lowering effect of P113 in patients with various forms of hypertension, do you think that there could be any reason why Dr. STREETEN seems to find some special blood pressure lowering effect in his patients with renal artery stenosis? Do you think that patients with renal artery stenosis do have something which, even when they have the same plasma-renin activity, makes them more renin-dependent than other patients?

ROBERTSON: I think that the experimental design is very different. We were studying people always recumbent, and nearly always on a fixed, regular, sodium-potassium diet for several days, whereas, if I remember Dr. STREETEN correctly, he had shortly before given the patients frusemide and had had them ambulant. So I think that it would be wrong to make a direct comparison – I think that they are two different styles of experiment. Even so, I do not think that our results were grossly incompatible.

BLAIR-WEST: Dr. ROBERTSON, on the dose-response curves for normals, it seems that the highest plasma aldosterone value that you had with infused angiotensin is 25 ng/100 ml. In a sodium-deficient patient, the lowest level there was 50 ng/100 ml, for roughly the same blood angiotensin II concentration. That is an interesting difference, but that is not my point. The point is that when you then infused saralasin into a normal, sodium-deplete subject, I noticed that the blood aldosterone concentration fell from say 35 ng/100 ml to about 25 or perhaps 20 ng/100 ml. Well, that is still a long way from basal aldosterone secretion rate. What is the reason that you do not get it back into the normal range? Because there is some other factor sustaining it? Or, because you are not giving enough saralasin, or is there an agonist action of the saralasin that prevents it falling?

ROBERTSON: Dr. OELKERS and our group have been interested for a long time in what factors may cause the enhanced aldosterone response to angiotensin during

---

1 Discussion of papers of BROWN *et al.* and FAGARD *et al.*

sodium depletion. Given that there is an altered response, one should not necessarily expect, if you infuse saralasin in sodium depletion, that you take yourself all the way back to where you were before you were sodium-depleted. For example, one of the lines which OELKERS and ourselves were pursuing was whether prolonged exposure to angiotensin II had some kind of hypertrophic effect on the adrenal gland. There was some evidence that we published in Circulation Research [36–37: suppl. 1, pp. 49–56, 1975] which suggested that that might be part of the explanation. The mean plasma aldosterone in our group of normal subjects in the present series, sodium-depleted, was 28.5 ng/100 ml before saralasin, 14.5 ng/100 ml during saralasin infusion, and returned to a mean of 36.5 ng/100 ml after stopping the saralasin. The upper limit of our normal range is 18 ng/100 ml.

GIESE: I would like to ask Dr. ROBERTSON to provide a little more information on the Glasgow experience with cross-reactions between angio-II and saralasin, because if I remember Dr. PETTINGER's figures correctly, you would have a 10,000-fold difference of plasma concentration when you infused 10 μg/kg/min, so there just might be a chance that even a slight cross-reaction could influence post-infusion results for angio-II in plasma.

ROBERTSON: Well, that's a very valid point. I rather brushed past this in the interests of getting at the chicken and champagne. We originally published on this showing that there was no detectable cross-reaction in our system, with up to 10,000 pg/ml of saralasin, and then we had another look at Dr. PETTINGER's figures, and did some calculations and worked out that during the high rate of saralasin infusion, 10 μg/kg/min, plasma concentrations would be about 20,000 pg/ml, at which level we did demonstrate some cross-reaction, equivalent to about 20 pg/ml of angiotensin II. However, there was no evidence of this effect after stopping saralasin; 1 h later the values were similar to those before the infusion.

MACGREGOR: I would like to ask Dr. ROBERTSON if he is confident about measuring the plasma level of angiotensin II after an infusion of angiotensin II and comparing it to an endogenous level of angiotensin II. The concentration around the receptor site may be quite different.

ROBERTSON: I think what you are getting at is an old idea of ours, which we have been hooked on for a long time. We have suggested that there may be a direct lymphatic vascular link between the adrenal gland and kidney – if you're getting at that, then, of course, I can't directly refute it. All I could say is that when we do infuse angiotensin, we seem to be on the same smooth dose-response curve of which the basal values are part. Dr. OELKERS is anxious to join in; he did some of the basic work on this when he was in our laboratory, and has continued it.

OELKERS: I think the only answer I could give is that the blood pressure is stable after 5 min of infusion, and we did all these infusions for 1 h; the aldosterone rises, reaches a plateau after half an hour, and we did not get any delayed change. Since angiotensin is infused into a vein in these experiments, while the endogenous angiotensin is formed throughout the circulation and perhaps in addition in blood vessel walls, angiotensin concentrations at the receptor site may be different in the two conditions. However, we measured angiotensin II in arterial plasma, which reaches the receptor within seconds.

GROSS: Thank you very much, indeed, and I think this brings us almost to the end. I should only like to say a word of caution, because the saralasin story (and not only that of saralasin, but also that of the other angiotensin II antagonists) is not nearly as easy as it looks. I am always intrigued by the fact that angiotensin begins to play a role in the regulation of blood pressure in the state of sodium depletion, when the responsiveness to angiotensin is reduced; this is in some ways strange. It makes much more sense in the case of aldosterone regulation, and I am most grateful that Dr. ROBERTSON reminded us of this with the dose-response curves he presented – because in the state of sodium depletion, the response of the adrenal gland to angiotensin is enhanced. But why, in a sodium-depleted state, when the dose-response curves are shifted to the right, should angiotensin contribute to the regulation of blood pressure? This is strange, to say the least.

Subject Index

Acetylcholine 118
ACTH 42, 44, 45
– release 2, 5, 7, 46, 118
ADH secretion 117, 119, 122–124, 130, 131, 143
Adrenal blood flow 54, 56, 85
– cortex, see Zona
– transplant 54, 61
Adrenalectomy 18, 19, 75, 99, 223
Adrenaline 215
Adrenergic blockade, α- and β- 171, 172, 204, 210
Affinity 136
Agonistic activity, see Angiotensin analogues
Aldosterone biosynthesis 7, 8, 41–47, 53
–, plasma concentration, see Plasma
– secretion, in dog 2–4, 99
– –, in man 206, 208–211, 247
– –, in rat 4–9
– –, in sheep 41, 53–61
see also Sar¹-Ala⁸-angiotensin II
Amino acids 136
Amino-peptidase 51, 101, 109, 115
Angiotensin analogues, see Sar¹-Ala⁸-, Sar¹-Ile⁸-, and Sar¹-Thr⁸-angiotensin II
– –, comparative studies 33–39
– –, limitations 34
– assay, see Radioimmunoassay

– dependence 114, 115, 158, 159, 171, 172, 185, 191–198, 210, 211, 214–224, 230–240
see also Hypertension
–, dose-response curves 50, 84, 115, 252
–, heptapeptide (angiotensin III) 8, 9, 19, 41–47, 50–52, 57, 58, 61, 207
–, hexapeptide 41–47
Angiotensin I, conversion in vessel walls 98–111
–, immunization against 105
– in brain 119, 129
–, in vitro studies 41–47, 136
– infusion 103
–, intrinsic activity 46, 50
– perfusion 99
Angiotensin II analogues, see Angiotensin
– antibodies and antisera 64, 98–111, 115, 116
– blockade, see Angiotensin II inhibitors
–, central actions 117, 134
– – formation 120–124, 129–131
–, dipsogenic effects 46, 117, 120, 121, 124, 142–144
–, effect on ADH, see ADH
–, effect on aldosterone 54
–, endogenous 187
–, free-circulating 102, 107, 116
–, half-life 109
– in cerebrospinal fluid 124, 129, 143

Subject Index

–, *in vitro* studies 41–47, 109
– inhibitors, *see* Angiotensin analogues, Converting enzyme inhibitor
–, intracerebroventricular injection 128
–, intrarenal formation 73, 74, 76
– – infusion 66–77
–, intravenous infusion 7, 58, 59, 61, 101, 121, 128, 215
–, local generation 98–111
–, metabolic clearance 58
–, myotropic responses to 33
–, plasma concentration, *see* Plasma
–, pressor responses to 104, 105, 115
– receptors 8, 16, 41–47, 100–109, 118, 128, 158, 183, 184, 251
–, relation to aldosterone 238, 239
–, role in hypertension, *see* Angiotensin dependence
Angiotensinase 115, 207
Antibodies, *see* Angiotensin II, Renin
Antidiuresis 123
  see also ADH
Antihypertensive drugs 197, 198, 209–212, 219
  see also Bethanidine, Diazoxide, Hydrallazine, Methyldopa, Propranolol
Anuria, *see* Renal failure
Aortic aneurysm 223
– –, caval fistula 9–14
Area postrema 118
Arterial blood pressure 2–4, 7, 11–14, 17–29, 35–39, 88–94, 100–111, 120, 121, 124–131, 146–248
– – –, direct recording 34, 243
– – –, dependence on angiotensin, *see* Angiotensin
– – – – on renin, *see* Renin
– – –, regulation 145, 163, 168, 169, 204
– – – relation to plasma angiotensin 236, 237
– – – – – – renin 181–183, 245–248
Arteriosonde 146, 191, 217
Arteriovenous fistula 73
  see also Aortic aneurysm, caval fistula
Ascites 163–169, 176–187
Asp^1-Ile^5-AII, *see* Angiotensin II

Assay, *see* Radioimmunoassay
– by protein binding 42
Autoregulation, renal 74

Bartter's syndrome 176, 178, 180–187
Bethanidine 233
Blood-brain barrier 118, 129, 130
– pressure, *see* Arterial blood pressure
– urea 195
– volume 170, 185, 209
  see also Vascular volume
Body weight and dialysis 190–198, 201
– – with sodium depletion 173, 247
Brain, renin-angiotensin system 117–131
–, renin substrate, *see* Renin
Buffer, bicarbonate 42

Carboxypeptidase A 136
Cardiac failure, *see* Heart
– output 92, 243–248
Catecholamines, secretion 34, 39, 46, 91, 151, 158, 175, 215
Cathepsin D 118, 130
Caval constriction 9–14, 73, 75, 127, 184
Cellophane perinephritis 88
Central cholinergic nerves 118
Cerebrospinal fluid 119, 129
Chlorthalidone 146–148, 161, 217, 243
Chromatography, paper 42
Chymotrypsin 136
Cirrhosis 163–169, 176, 178, 180, 182–187
Collateral vessels 88, 114
Competitive inhibitors, *see* Renin
Converting enzyme 39, 99, 119, 124, 125, 131, 207
– – inhibitor 6, 7, 13, 14, 16–29, 45–52, 73–75, 99, 103, 105, 110, 111, 119, 121, 124, 126, 127, 129, 159
Corticosterone 3, 5–8, 42–47
Cortisol 4, 5, 42–47, 60
Cortisone 18
Creatinine clearance 11, 222, 243
Cross-circulation 100
Cushing's syndrome 176, 178, 179, 185, 223, 224
Cumulative sodium balance 148–153, 157, 160

Subject Index

Des-Asp[1]-angiotensin II, *see* Angiotensin heptapeptide
Dexamethasone 2, 5–8, 35, 51, 53, 186
Dialysis machine, *see* Haemodialysis
–, peritoneal 104, 105
Diazoxide 108, 195, 248
Dichloromethane 177
Dietary sodium, *see* Sodium
Dissociation constant 44
Diuretics 21, 151, 157, 165, 167, 175–177, 209, 212, 247
  see also Chlorthalidone, Furosemide, Hydrochlorothiazide, Spironolactone
DOCA (desoxycorticosterone acetate) 18, 51
Dogs 86–95, 119–131, 186
Drinking, regulation 119–127
  see also Angiotensin II

Edetate (EDTA) 177, 191
Effective blood volume, *see* Blood
Electrolyte disorders 174–187, 234, 235
Epinephrine, *see* Adrenaline
Essential hypertension 95, 145–161, 176, 178, 182–187, 233
– –, blood pressure responses to Sar[1]-Ile[8]-angiotensin II 174–187
– – – – – to saralasin 152–161, 190–198, 204–212, 214–224
– –, renin dependence 159, 160
– – – subgroups 153, 159, 178, 182–187, 210, 211
– –, saralasin and sodium balance 145–161
Ether anaesthesia 65
Experimental hypertension, *see* Hypertension, Renal hypertension
Extracellular fluid volume 18, 94

Fainting 146–148
Feedback, *see* Negative feedback
Fluid disorders 174–187, 190–198
  see also Ascites
Free water clearance 122, 123
Furosemide 27, 128, 176–178, 216–224, 227, 233

Glomerular filtration rate 247
  see also Creatinine clearance

Glomerulonephritis 192, 193
Goldblatt hypertension, *see* Renal
Glycerol, in producing renal failure 65–70

Haemodialysis 190–201, 233–235
Haemodynamic studies 242–248
Haemorrhage 75
Heart failure, congestive 99, 176, 178, 180–187
– –, high output 9–14, 75
– –, low output 4, 10, 29, 195
– rate 49, 171, 201, 245–248
Hydrallazine 193–196
Hydrochlorothiazide 99
8-Hydroxyquinoline 65, 191
Hypertension, angiotensin-dependent (angiotensinogenic) 90, 93, 200, 201, 208, 214–224, 230–240, 242–248
–, clinical, *see* Cushing's syndrome, Essential hypertension, Phaeochromocytoma, Primary aldosteronism, Renal hypertension
–, experimental 86–95, 101, 104
  see also Renal hypertension
– in dialysis patients 190–198, 233–235
–, malignant 12, 75, 88, 91, 95, 129, 159, 176–179, 182–187, 195, 231
–, renin-dependent 26–29, 159–161
–, renovascular, *see* Renal hypertension
–, volume-dependent 190–198, 247
Hypophysectomy 2, 3, 6, 7, 123, 223
Hypotension, orthostatic 164, 166, 193, 197
  see also Fainting
–, severe 27, 28, 201, 218, 220
Hypovolaemia 126, 158, 184–186

Immunization against angiotensin II 76, 93
– with renin 25
Indomethacin 75
Inhibitor, *see* Angiotensin II, Renin
Isolated perfused kidney 63–77
Isoproterenol 127
Isorenin 50
Intrinsic activity, *see* Agonistic activity, Angiotensin I

Juxtaglomerular apparatus 2, 50, 73, 224

Subject Index

K_i 136–139, 144
K_m (Michaelis constant) 44, 116, 137–139, 144
Kidney, see Isolated perfused kidney, Transplant
Krebs-Henseleit solution 64

Leucine 136, 137
Lima bean 42
Lineweaver-Burk plots 44, 136
Logit-log plots 43

Macula densa 74
Malignant hypertension, see Hypertension
Me_2Gly^1-Ile^8-AII 74
Meclofenamate 75
Mean blood pressure 151, 154, 176, 177
Mesenteric artery 99, 100
Methyldopa 192, 193
Mineralocorticoids 42–47, 185
  see also Aldosterone, Corticosterone, DOCA
Morphine 6–8

Natriuresis 218, 219
Necrotizing vascular lesions 91
Negative feedback of angiotensin II on renin release 2, 19, 22, 26, 29, 34, 54, 59, 75, 107, 160, 175, 186, 206–210, 228, 229
– –, tubulo-glomerular 74
Nembutal (sodium pentobarbitone) 64, 121–123, 128
Nephrectomy, bilateral, dog 1, 2, 10, 99, 100, 186
– –, man 192, 196, 201, 215, 217, 223
– –, rat 100–111, 129
– –, sheep 53
–, unilateral, dog 22, 88
– –, man 223
Nephrosclerosis 95
Nephrotic syndrome 176, 178, 192
Noradrenaline (norepinephrine) 76, 88, 100, 118, 215
Normal (normotensive) subjects 146–152, 157–161, 163–167, 176–187, 191, 219, 233

Oedematous states 164–169, 174–187, 217
Optic fundus 243

Osaka, sodium intake in 184
Osmometer 65
Osmolality, urine 65
Osmolar clearance 122, 123
Oxygen content 244
–, Fick method 243
– uptake 243

P113, see Sar^1-Ala^8-angiotensin II
Papain 136
Papilloedema 195
Partial agonists, of AII 71
Pepstatin 119, 121, 124, 129, 140
Peptide inhibitors of renin 135–140
–, effects of pH 136–140
– solubility 136–138
Perinephritic hypertension 87–95
Peripheral resistance 92, 103, 246–248
Phaeochromocytoma 176, 178, 182, 183, 185, 201, 202, 215
Phenoxybenzamine 204, 215
Phentolamine 204, 215
Phenylalanine 33, 45, 137
Phenylephrine 18, 19, 49
Phenylmethylsulphonylfluoride 191
Pituitary gland 118
Placebo 21
Plasma, ADH concentration 123
–, angiotensin II concentration 24, 57–59, 61, 73, 110, 191–198, 208, 211, 215, 232–240, 250, 251
–, aldosterone concentration 2, 8, 14, 20–22, 27, 28, 35, 36, 53, 165–168, 174–187, 215, 230–240, 250
–, potassium concentration 35, 53, 60, 85, 195
–, renin activity (PRA) 2, 4, 9, 11–14, 17, 18, 20–24, 27, 28, 34–39, 49, 72, 73, 88–95, 99, 101, 104, 137, 147, 154, 157, 165–168, 174–202, 208, 209, 215–217, 220–224
– – concentration 54–60, 72–75, 233–240, 243–247
– –, in renal vein 26, 27
–, sodium concentration 35, 60
–, urea concentration 70, 76, 77
Postural changes, effects on aldosterone 22

– – with saralasin 152–161
Potassium balance 171
–, dietary 146, 147
– in plasma, see Plasma
–, urinary 123
Pressor activity of angiotensin analogues, see Angiotensin
–, responsiveness to angiotensin II 102–105
Pressure transducer (Statham) 64
Primary aldosteronism 176, 178, 179, 182, 186, 211, 215, 234, 235
Progesterone 99
Propranolol 18, 74, 193–196, 200, 201, 206, 208, 209, 211, 223
Proline 136, 137
Prostaglandins 75, 100, 216
Protein-binding assay 42
Pulse rate 164, 167
Pump, infusion 88, 191, 231
Pyelography 147, 222, 243
Pyelonephritis 28

Radioimmunoassay, aldosterone 35, 165, 177
–, angiotensins 42, 45, 149, 191, 230
–, renin activity 35, 65, 88, 149, 165, 177, 191
–, saralasin 204
Rat, pithed 101
– serum 68, 70, 71, 76, 77
–, spontaneously hypertensive 129
–, Sprague-Dawley 63–77
Receptors for angiotensin, adrenal, cortex 41–47, 84, 251
– – – –, glomerulosa 8, 16
– – –, brain 118, 128
– – –, vascular 16, 46, 100–111, 158, 183, 184, 208
Renal arteriography 147, 222, 227, 243, 244
– artery, aneurysm 244, 245
– –, constriction 26, 75, 87, 88, 114
– –, stenosis 208–212, 215, 216, 223, 224, 237, 244, 245, 248, 250
– –, surgery 222, 223
– blood flow 9–12, 72, 73, 76, 222, 247
– failure, acute 65, 67, 75, 76, 195

– –, chronic 176, 178, 179, 190–198, 223, 234, 235
– hypertension, clinical 26–29, 159, 176, 178, 187, 190–198, 208, 214–224, 230–240, 242–250
– –, experimental 9–14, 22–26, 87–95
– –, 'one kidney' 13–26, 29, 87–95, 159, 216
– –, perinephritic 87–95
– –, sodium retention 25, 26, 176–181, 216, 224
– –, 'two-kidney' 13, 14, 72, 87–95, 100, 105–116, 159, 216, 247
– ischaemia 99
– perfusate flow 66, 68, 69, 73
– prostaglandins 75
– transplant 192, 193
– tubular acidosis 176
– vascular resistance 10, 72–77
– vasoconstriction 66, 67, 72, 73, 75, 76
– vasodilation 72, 73
– vein renin 26, 27, 208, 209, 216, 220–224, 228, 243, 244
Renin activity, see Plasma
– affinity chromatography 140
– antibodies 25
– assay 65, 100
see also Radioimmunoassay
– circulating 58, 104
– dependence 26–29, 114, 115, 159–169, 216, 250
– depletion 74
–, dipsogenic effects 119–121
– in adrenal glands 60
– in blood vessels 98–111
– in plasma, see Plasma
–, intracerebroventricular administration 119–123, 126, 127, 130, 131, 144
–, intravenous administration 101, 105–110, 120, 121, 123
–, peptide inhibitors 135–140, 143, 144
–, pH optimum 118, 143, 144
–, primate 138, 139
– purification 140
–, release (secretion), control 18, 19, 49, 53, 66–69, 72, 74, 75, 77, 104, 105, 107, 152,

Subject Index

158–160, 175, 186, 187, 197, 198, 206–212, 227–229
–, species specificity 138, 139
– subgroups, *see* Essential hypertension
– substrates 64, 65, 73, 74, 118, 119, 135–140, 143, 144, 185
– –, in brain tissue 124, 131
– –, in cerebrospinal fluid 119, 124, 142
– –, synthetic tetradecapeptide 124, 125, 135–139, 142
–, vascular 104–111
Renin-like activity 98–100
– –, in brain 118–125
– –, pH optimum 118–125
Retinopathy 147

Sanborn recorder 34
Sar^1-Ala^8-angiotensin II (saralasin, P113) 2–14, 17, 18, 25, 33–77, 94, 106, 109–131, 145–175, 184, 190–248
–, agonistic activity 34–39, 44, 46, 49–52, 59, 60, 64, 66, 71, 72, 110, 123, 154, 158, 160, 161, 168, 175, 184, 192, 200–202, 206, 218, 219, 234, 247, 250
–, catecholamine release 49, 151, 158, 175
–, clinical diagnostic use 145–161, 190–240
–, cross reactivity with AII 45, 233, 234, 251
–, effects on aldosterone 49–52, 163–172, 206–211, 233–240, 250, 251
– – on arterial blood pressure 3–5, 11–14, 35–39, 106, 109–111, 145–161, 163–169, 175, 190–198, 205–212, 214–224, 233–240, 245–248
– – on cardiac output 244–248
– – on corticosteroid output of fasciculata 41–47
– – on creatinine clearance 11
– – on filtration fraction 11, 73
– – on glomerular filtration rate 72, 73
– – on heart rate 49, 171, 201, 245–248
– – on renal blood flow 10–12, 72, 73, 76
– – on renin release 85, 152, 156–157, 163–169, 208
    see also Negative feedback
– – on relations between AII, aldosterone and blood pressure 236–240

– – on responses to ambulation and posture 151, 155, 163–172, 193, 197, 209
– – on responses to altered sodium balance 145–161, 163–173, 175–179, 228
– – on sodium excretion 11, 72, 73, 167
–, dosage 142, 143, 200, 219, 233–238, 243, 251
–, false test results with 211, 222, 224, 227, 228
–, half-life 52, 205
– in cirrhosis with ascites 163–169
–, *in vitro* studies 41–47
–, intraadrenal arterial infusion 53–61
–, intracerebroventricular administration 120–131, 142, 143
–, intrarenal arterial infusion 11, 53, 55, 72
–, intravenous infusion 67, 72, 106, 107, 146–161, 163–169, 190–198, 204–212, 214–224, 230–240, 242–248
–, pharmacokinetics 205, 206
–, plasma concentration 59, 204–207, 251
–, radioimmunoassay 45
–, receptor binding affinity 72
Sar^1-Gly^8-angiotensin II 74
Sar^1-Ile^8-angiotensin II 25, 33–39, 86–95, 174–189
–, agonistic activity 175, 183–187
–, effects on adrenal cortex 38, 39, 175
– – on arterial blood pressure 35–39, 89–95, 129, 174–187
– – on catecholamines 91, 185, 201
– – on plasma aldosterone 181–187
– in clinical disorders 174–187
– in perinephritic hypertension 91
– in renovascular hypertension 88–95
–, intravenous infusion 174–187
–, synthesis 176
Sar^1-Thr^8-angiotensin II, agonistic activity 93
–, effects on adrenal cortex 38, 39, 51, 52, 93
– – on arterial blood pressure 35–39, 89–95, 129
– in perinephritic hypertension 91
– in renovascular hypertension 88–95
Sepharose 139

Subject Index

Serum creatinine 147, 222
SQ 20881, *see* Converting enzyme
Snake venom 16, 17
Sodium and water retention 25, 26, 176–181, 216, 224
– balance 17–20, 101–105, 145–161, 172–175, 183, 204, 239
– chloride 74, 98, 218
– depletion, dog 1–3, 13, 14, 17, 18, 35–37, 51, 72, 73, 75, 127–129
– –, man 20, 22, 27–29, 51, 146–161, 163–173, 175, 184, 192, 197–201, 217, 218, 227, 231, 233, 236–239, 242–252
– –, rabbit 73
– –, rat 4–7, 75, 94, 101, 104, 105
– –, sheep 53–61, 84
–, dietary 146, 163–173, 175–178, 181, 184, 191, 217, 218, 243
–, handling, defect in 159
–, insensible losses 149
– loading 51, 101, 105, 158, 176–181, 190, 218, 222, 224, 228
– pentobarbitone (Nembutal) 64, 121–123, 128
– -pressor factor 160, 161
–, renal tubular reabsorption 12
Solute excretion in anuria 71, 76, 77
Spironolactone 147, 161, 215
Spontaneously hypertensive rats 129, 131
Statistical methods 35, 65, 183, 244
Steroids, adrenal, *see* Aldosterone, Corticosterone, Cortisol
Stroke volume 246

Swan-Ganz catheter 246

Tachycardia 20, 22, 171
Tetradecapeptide substrate, *see* Renin
Tilting stress 20, 21, 220
Transplant, renal, *see* Renal
Trypsin 42
Tubulo-glomerular feedback 74
Tyrosine 137

Urease 65
Urinary electrolytes 123, 149
– osmolality 122, 123
Urine flow 122–124
– volume, in anuria 70, 71, 76, 77

V_{max} 44
Vascular reactivity 104, 105, 115, 185
– volume 170, 207, 209, 247
– –, increased 184–186
– –, reduced, *see* Hypovolaemia
Vasodilator drugs 193–196, 210, 211
Vasopressin 76, 142, 143, 205

Water deprivation 126
– diuresis 122
– intake 131
– loading 190–198

Xanthine derivatives 74

Zona fasciculata 4, 41–47
– glomerulosa 8, 45, 187
– reticularis 4

157095
v.12-